CW00927341

Military Architecture

The Tower of London in the Fifteenth Century.
(From *London*, Edited by Charles Knight, 1841)

Military Architecture

by

E. E. Viollet-le-Duc

Greenhill Books, London
Presidio Press, California

This edition of *Military Architecture*
first published 1990 by Greenhill Books, Lionel Leventhal Limited,
Park House, 1 Russell Gardens, London NW11 9NN
and
Presidio Press,
31 Pamaron Way, Novato, Ca.94949, U.S.A.

Lionel Leventhal
Celebrating 30 years of military publishing
1960–1990

British Library Cataloguing in Publication Data
Viollet-le-Duc, Eugene Emmanuel *1814–1879*
Military Architecture.
1. Military Operations, History
I. Title II. Macdermott, M.
355.409

ISBN 1-85367-078-2

Publishing History
Military Architecture was first published in English in 1860, translated from the French by M. Macdermott. This edition reproduces the Third Edition (James Parker and Co., 1907), complete and unabridged, with some additional illustrations.

Printed in Great Britain by
Billing & Sons Ltd, Worcester

PREFACE TO THE SECOND EDITION.

THE first edition of the English translation of this work was published in 1860, under my direction, with the full consent of the Author, and with the original engravings from his own excellent drawings. My reason for re-publishing it at the present time is because I cannot help seeing how useful it would be for the officers of the English army in Zulu-land and other parts of South Africa, and in the savage parts of India, wherever the well-disciplined troops of civilized nations come in contact with savages. It explains all the modes of attacking and defending a camp or a city that have been used from the time of the Romans, by means of these admirable drawings of M. Viollet-le-Duc, which make them far more easy to understand than any words alone could do. It is much to be regretted that more attention has not been paid to this subject in military schools. The first twenty pages of this work, describing the Roman fortifications, would alone be of immense service in fighting with barbarians, as they had to do. The Roman remains in many parts of England would be easily accessible, and much might be learned from them. Unfortunately the fortifications of the ancient Britons on the hills are continually called by ignorant people "Cæsar's Camps;" and the earthworks of the real Roman camps, which may generally be found within half-a-mile of the British city, are entirely overlooked, and allowed to be ploughed-up by the farmers. In some parts of England the entrenchments of many cities and camps remain, and they are part of the history of the country and of the people.

When the Romans first attempted the conquest of Britain under Julius Cæsar, they attacked these fortified cities by assault, but although their superior arms and discipline generally enabled them to take them, their commanders soon saw that this cost too many lives; half the Roman army was lost in a single campaign, and they were obliged to retire and abandon the attack for a time, but never lost sight of their object; and when they resumed the attack in the time of Claudius and Nero, under the great general Aulus Plautius Lateranus, they had learned caution, and they no longer attacked by direct assault, but starved out the Britons by blockade, especially by cutting off their supply of water. The cities were almost always on the tops of lofty hills, and strongly defended by entrenchments, following the outline of the cliffs of the hill; but at that height there was no water, and that necessity of life had to be fetched from some neighbouring stream or spring on the level ground below. The Romans, therefore, made a camp at a short distance; their entrenchments were always on the oblong plan, and on level ground; a Roman legion could entrench their camp in a single night sufficiently to be secure against the attack of savages; then they followed their usual practice, one-third of their army slept in turn, another third was ready to defend the camp, and the remaining third to attack or to watch the enemy. The first object was to watch where the water was fetched from, and then place a guard upon that day and night, relieved in turn, so that each had to watch for eight hours only, by that means they could entirely prevent the Britons from getting any water. In their camps, their superior arms and discipline could not be overcome by their half-armed and half-naked assailants in whatever numbers they might surround them; and the Romans always made a well in each camp, unless it was by the side of a running stream, as at Dorchester, in Oxfordshire.

It appears that even before their time the Etruscan soldiers understood equally well the importance of the spade in war; there is in the Kircherian Museum in Rome a figure of an Etruscan soldier, in bronze, with a wheel-barrow fastened to his back instead of the knap-sack of an English soldier, and apparently not much more difficult to carry.

Etruscan Soldier with Wheelbarrow, from the Kircherian Museum, Rome.

This is a very curious illustration of the *defensive arms* (if they may be so called) of that early period. This man probably belonged to the corps of engineers; we see that he has the shield and his wheelbarrow, but neither sword nor spear. It seems evident that a certain number of these trained and skilled workmen, thus equipped, accompanied the army, and that their duty was *to dig* and form an entrenchment at once, and defend themselves in case of need while doing so, not to fight. In such a country as South Africa might not the same plan be followed with advantage? Probably the natives would be more ready to dig than to fight for the stranger, and might safely be trusted to do the digging part, which would set at liberty to fight so many more of the English soldiers, and relieve them from some of the hardest work in that hot climate. All accounts agree that the cities of the Zulus are placed in the same manner, on the tops of inaccessible mountains, and there could very rarely be any water in that situation; could they not be watched in the same manner as the Britons were? It has long been known that the Roman soldiers conquered more by the spade than by the sword from the earliest period of their history. The original fortifications of Rome itself were entirely earthworks, *aggeres*, or great banks of earth, with enormous *fossæ*, or trenches; the lesson they had learned at home they carried with them everywhere.

It may seem strange, at first sight, for an army in the nineteenth century to take a lesson from the tactics of warriors of a thousand years before the Christian era; but it must be remembered that the people we have to fight with are in much the same state of civilization as the natives of Italy were when they were conquered by the Etruscans, and that EARTHWORKS are found by experience to be *the best defence* against modern artillery, as they were at an earlier period against the artillery of those days.

At a later period the Norman Barons, to whom large grants of land in all parts of England were made by William the Conqueror, had to defend themselves against a hostile population, and each built himself a castle; at first these were earthworks only, but "necessity, the mother of invention," taught Gundulph to erect a keep of stone on his estate at Malling, in Kent. This keep is in existence, and is of the rudest construction; M. Arcisse de Caumont and the best Norman antiquaries acknowledged that it is of earlier character than any keep in Normandy, when I shewed them a drawing of it, and the French call the style of architecture which we call THE NORMAN STYLE, the ANGLO-NORMAN STYLE. There is no doubt that it was developed in England under the Norman kings quite as much as in Normandy. About the year 1840 M. Arcisse de Caumont, the leader of the Norman Archæological Society, with a party of the best-informed members of it, made an excursion to the sites of all the castles of the Norman Barons who had gone over to England with William the Conqueror, with the intention of ascertaining the distinction between the *construction of walls* in the first half of the eleventh century and those of a later period. To their astonishment they could find no trace of *masonry* of that period in any one of them. There were magnificent EARTHWORKS in all of them, but no stone walls. An account of this excursion is given in the *Bulletin Monumental*, of which M. de Caumont was the editor.

There is good reason to believe that Gundulph was *the inventor* of the Norman keep; it was so exactly what was wanted, that it was rapidly followed all over England, and gradually spread over the whole of Europe, especially wherever the Normans went. There are Norman keeps in Sicily, and in the south of Italy. Gundulph became a great favourite of the Conqueror, was rapidly promoted, and was soon made Bishop of Rochester, where he built another keep, of which also there are remains of rude work, almost as early as the one at Malling (which is miscalled S. Leonard's tower). The one at Rochester is not on the site of the present castle (which is half a century later in date), but forms a sort of transept to the present cathedral.

M. de Caumont was the first person to introduce the study of ARCHITECTURAL HISTORY into France; he always candidly acknowledged that he took the idea from Rickman's work, which was first published in 1810, and the third edition in 1830; it was not till the later date that the subject had attracted much attention in England. Rickman was the first "to reduce chaos into order," but it was Professor Willis, in his admirable history of Canterbury Cathedral, in 1844, who fully developed the idea, and for some years after this, in his lectures on the other cathedrals delivered also to the Archæological Institute, he explained the system to all the leading architects of the time, including Sir Gilbert Scott, who always acknowledged how much he owed to the guidance of Willis.

Again, in the thirteenth century, during the long reign of Henry III., the wild Welshmen were continually making raids into England, and driving off herds of cattle, then the chief wealth of the country; much in the same way as the savage Zulus were expected to make raids in the British colony of Natal, which is perhaps about as civilized as England was in the time of Henry the Third. There was then no standing army in England, and so large a proportion of the English people were engaged in building that they were not prepared to drive back these Welsh hordes; but on the accession of Edward I. to power, he, being the most able monarch that England has ever seen, at once saw the necessity of keeping the Welsh thoroughly in subjection, and to do this he established a chain of English settlements all round the border of Wales; each of these settlements consisted of a castle and a fortified town, the castle was independent of the town, and could be defended even against the citizens in case of need, but this was not likely to be the case, as the main object of the castle was to protect the town, and to receive the citizens in case they should be overpowered. These English settlements were for the most part near the mouth of a river and the sea-coast, so that they could be supplied with provisions by sea in case of need, but would usually obtain them from their Welsh neighbours, who voluntarily took their produce to market, the rivers being the chief highways at that period. The measures of this wise monarch were perfectly successful, and there were no more raids of the wild Welshmen into England.

M. Le-Duc's admirable views of the fortifications of Carcassonne are a thoroughly good illustration of the fortifications of that period. Why cannot the commanders of the English forces in Zulu-land take a lesson from experience? The earthworks of the Roman fashion would be perfectly sufficient for the present, stone walls may hereafter be necessary to protect the English towns; why should not these be placed in similar situations to the English towns on the border of Wales, where they may be supplied with provisions by sea in case of need, or in ordinary times may be well supplied by the natives by means of the rivers? If the present work had been a school-book in military colleges, as it ought to be, English officers would not now have this lesson to learn; but it is not too late—it is evident that the conquest of the Zulus must be a work of time, it is not a thing to be done in a day or a year; it took the English half-a-century to conquer the Welsh, let us hope it will not take them as long to conquer the Zulus, although they are very far off, and not close at hand as the Welsh were, neither are they such formidable enemies. The mountains and the fastnesses of Zulu-land are not worse than were those of Wales, let us hope that in the next half-century the natives and the English settlers will be as cordially united as the Welsh and the English are.

The only one of the English settlements in Wales in which the fortifications of the castle *and town* remain nearly perfect, is Conway, and the manner in which this important historical monument is neglected by the English Government is very disgraceful to it; there is no doubt that the walls are Crown property, the accounts for building them are still in the Public Record Office, and they were not sold to the Municipality by Pitt in the time of George the Third, as most of the fortifications of the English towns were, yet the inhabitants are allowed to make all sorts of encroachments upon them, and each to claim some portion of the old wall as his own private property; probably the next generation will find very little remaining, and so an important chapter in the history of England will be erased by the carelessness of the Government. It would be long before fortifications in Zulu-land will be considered as useless, and the natives allowed to destroy them at pleasure.

JOHN HENRY PARKER, C.B.

ADVERTISEMENT TO THE FIRST EDITION.

THE work now offered to English readers has already attained an European reputation in its original language. The accomplished Author has thrown entirely new light on an interesting subject, and has brought to bear upon it not only the results of his great experience as an architect, but also shews a thorough knowledge of the principles of engineering, and great research as an antiquary. The remains of our ancient castles will no longer be considered merely as picturesque ruins, but as objects of careful study, worthy of minute examination in order to discover not only the age when each part was built, but also the special purpose for which it was built with a view to the defence of the castle or town. That part of the work which relates to the hoarding, or wooden constructions to assist in the defence of the castle, is entirely new, and explains many things which were previously quite incomprehensible. The Author pays a just tribute to the memory of Richard Cœur-de-Lion, as not only a brave warrior, but also an accomplished engineer, in advance of his age, shewing great ability and skill in the construction of the Château-Gaillard; and this portion of the work cannot fail to be interesting to English readers. It also affords another instance of the valuable assistance which a knowledge of Medieval Architecture is calculated to render to History. A great deal of medieval history is hardly intelligible without it, and the successive changes in the modes of warfare as developed in this work explain many important passages, especially in the wars between France and England. We now see some of the causes why the English were successful at one period and the French at another.

CONTENTS.

	PAGE
INTRODUCTION	1, 2
The Visigoths in the Fifth Century	3
The Barbarians imitated the Romans	3
Wooden Ramparts of the Celts	4, 5
The Roman Testudo	6
The Roman Camps	7
Wooden Towers on Roman Walls	8
Roman Walls of Towns	9
The Visigoth Fortifications	10
Towers at Carcassonne	11
Tower with Outworks	12
Fortifications of Towns	13
Roman Towns	14, 15
Visigoth Towns	16
Roman Fortifications	17—22
The Tower	19, 20
The Rat	21
Attacks of the German Tribes	23, 24
The Battering-ram used in the Tenth Century	25
The Battering-ram used in the Eleventh Century	26
Improvements after the First Crusades	27
Detached Forts introduced in the Twelfth Century	28
Advantages of Detached Forts	29
Frequency of Sorties	30
The Norman Castles	31
Activity of Defenders Necessary	32
The Engines of War	33—36
The Mine	37
Siege of Carcassonne in 1240	38—42
The Battering-ram in the Thirteenth Century	43

PAGE

Siege of Toulouse by Simon de Montfort . . 44—46
Fortifications of Carcassonne . . . . . 47—59
Plan of Carcassonne 56
Bird's-eye View of Carcassonne 57
Necessity for Projections from the Walls . . . 60
The Hoarding 61
The Hoard, and the Cat 62
The Lines of Approach 63
Engines for Attack and Defence 64
Attack by the Drawbridge from the Wooden Tower . . 65
Use of Bastions 66
Defensive Arrangements 67
Details of Defence 68
Means of Defence 69
Spirit of Feudalism 70, 71
The Feudal Castle 72
Paris and the Louvre 73
Plan of Paris, Thirteenth Century 74
Plan of Paris, Fourteenth Century 75
Plan of Coucy 76
Plans of Towns 77
Anglo-Norman Feudalism 78
Feudal Castles of France 79
RICHARD CŒUR-DE-LION a consummate warrior and *an able engineer* 80
The Château-Gaillard . . . . . . 81—90
Keep of the Château-Gaillard 91
Siege of the Château-Gaillard by Philip Augustus, defended by Roger de Lacy . . . . . . 92—94
Castle of Montargis 95, 96
The Donjon or Keep . . . . . . 97—99
The Donjon or Keep of Etampes 99
The Donjon or Keep of Provins . . . . 100—104
The Castle and Keep of Coucy . . . . 105—113
Enguerrand de Coucy 114
The Feudal Castles 115
Improved Modes of Defence 116
Arrangement of Loopholes 117
Loopholes and Battlements 118
Round Bastions 119
The Curtain-wall 120

	PAGE
The Pointed Bastion or Horn	121
Pointed Bastions or Beaks	122
Bastions at Aigues-Mortes	123
Plans of Bastions at Carcassonne and Falaise	124, 125
The Narbonne Gate at Carcassonne	126—131
Means of Defence	130
The Drawbridge	132—134
Siege of Aubenton	135
Timber-hoarding	136—139
Battlements and Machicoulis	140
Hoarding and Machicoulis	141
Castle of Pierrefonds	142—144
The Walls of Avignon	145—149
Palace of the Pope at Avignon	150
The Castle of Vincennes	151
Plan of Vincennes	152
Improvement of Defences	153
Introduction of Infantry	154
The Battle of Crécy	154
Changes in Warfare	155—160
The Siege of Aiguillon	156
The Siege of Calais	157—159
The Jacquerie or Brigands	161, 162
Superior Discipline of the English Armies	161
The Army of Du Guesclin	163
Feudal Traditions long preserved, except in the Good Towns	164, 165
Introduction of Artillery	166
Early Use of Artillery and Trenches	167, 168
The English Expelled from France by Improved Artillery	169
Further Improvements in Artillery	170, 171
Cannons of the Fifteenth Century	172
An Archer of the Fifteenth Century	173
The Long-bow and the Cross-bow	174, 175
Alterations of Castles to receive Cannon	176, 177
The Castle of Bonaguil	178—180
Embrasures for Cannon	181
Modifications of Towers	182
Walls of the Town of Langres	183—186
Adaptations of Old Works	187
Tower at Perigueux	188, 189
Fortress of Schaffhausen	190—192

	PAGE
Fortifications of Schaffhausen	193—198
Changes in the Art of Defence	199—201
Fortifications of Orange under Louis XI.	200
Fortifications of Nuys	202
Castles of the Close of the Fifteenth Century	203
Modes of Strengthening Walls	204
Re-entering Ramparts	205
Fortifications of Sienna	206
Effects of Artillery	207
Use of Discharging-arches	208
Ramparts for Artillery	209
Ramparts of Earth and Timber	210
Ramparts of Timber	211
Embrasures formed with Gabions	212
The Trenches with Gabions	213
Fortifications of Metz	214, 216, 257, 258
Widening the Area	215
Enlargement of Barbicans into Boulevards	215
Fortifications of Hull	217, 218
Fortifications of Lubeck	219
Fortifications of Milan	220, 221
Use of the Cavalier	222
The Bridge of Marseilles	223
Cavalier at Verona	224
Use of Traverses	225
Use of Bastions	226
Fortifications of Nuremberg	227, 228
Fortifications of Augsburg	229—232
Frankfort-on-the-Maine	233, 234
The Orillon	235, 236
The Italian Engineers	237
Improved Bastions	238
Bastions attacked	239, 240
A Bastion Isolated	241
Bird's-eye View of a Bastion	242
Use of Ravelins	243
Plans of Ravelins	244
Improvements in Embrasures	245
Embrasures at Nuremberg	246, 251
Hoarding at Nuremberg	247
Improved Embrasures	248

PAGE

Crenelles with Shutters 249
Embrasure with Loopholes 250
Embrasures at Basle 252
Complicated Defences 253
Advice of Machiavelli 254
Changes caused by the use of Artillery 255
Effects of Artillery 256
Mines and Countermines 259
Countermines—Galleries 260
Bastions according to De Ville . . . . 261—263
System of Vauban 264
Conclusion 265

THE TOWER OF LONDON

From a Print published by the Royal Antiquarian Society, and engraved from the Survey made in 1597, by W. Haiward and J. Gascoigne, by order of Sir J. Peyton, Governor of the Tower. *a*. Lion's Tower; *b*. Bell Tower; *c*. Beauchamp Tower; *d*. The Chapel; *e*. Keep, called also Caesar's, or the White Tower; *f*. Jewel-house; *g*. Queen's Lodgings; *h*. Queen's Gallery and Garden; *i*. Lieutenant's Lodgings; *k*. Bloody Tower; *l*. St. Thomas's Tower (now Traitor's Gate); *m*. Place of Execution on Tower Hill.

(From *Mediaeval Military Architecture in England* by Geo. T. Clark, 1884)

THE TOWER OF LONDON

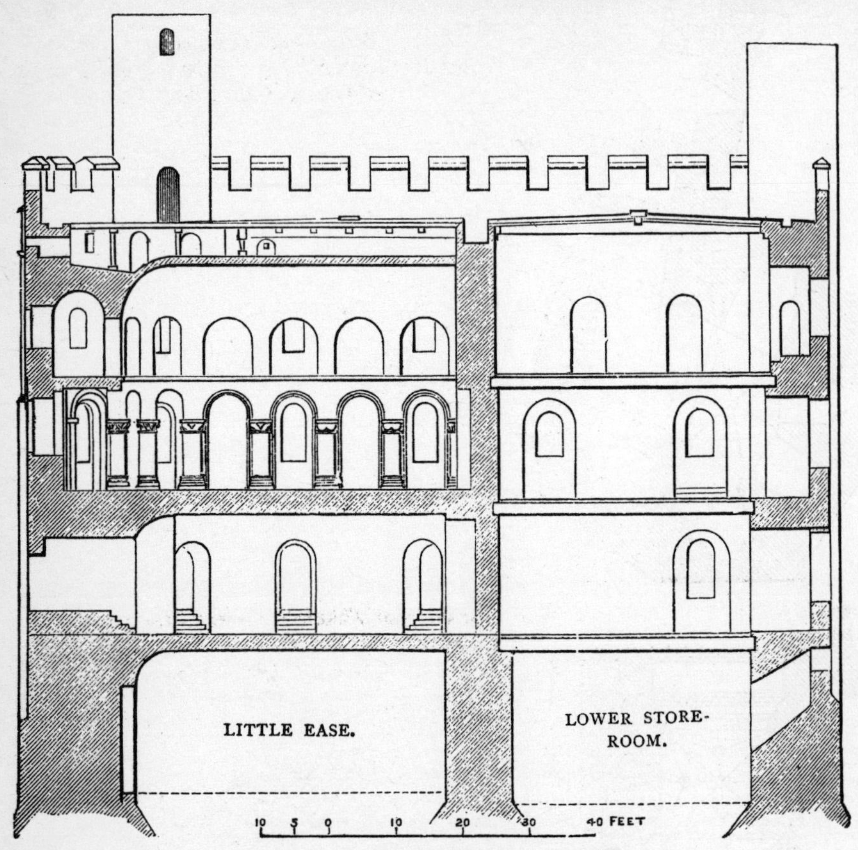

The Keep. Vertical Section, East and West.
(From *Mediaeval Military Architecture in England*
by Geo. T. Clark, 1884)

THE TOWER OF LONDON

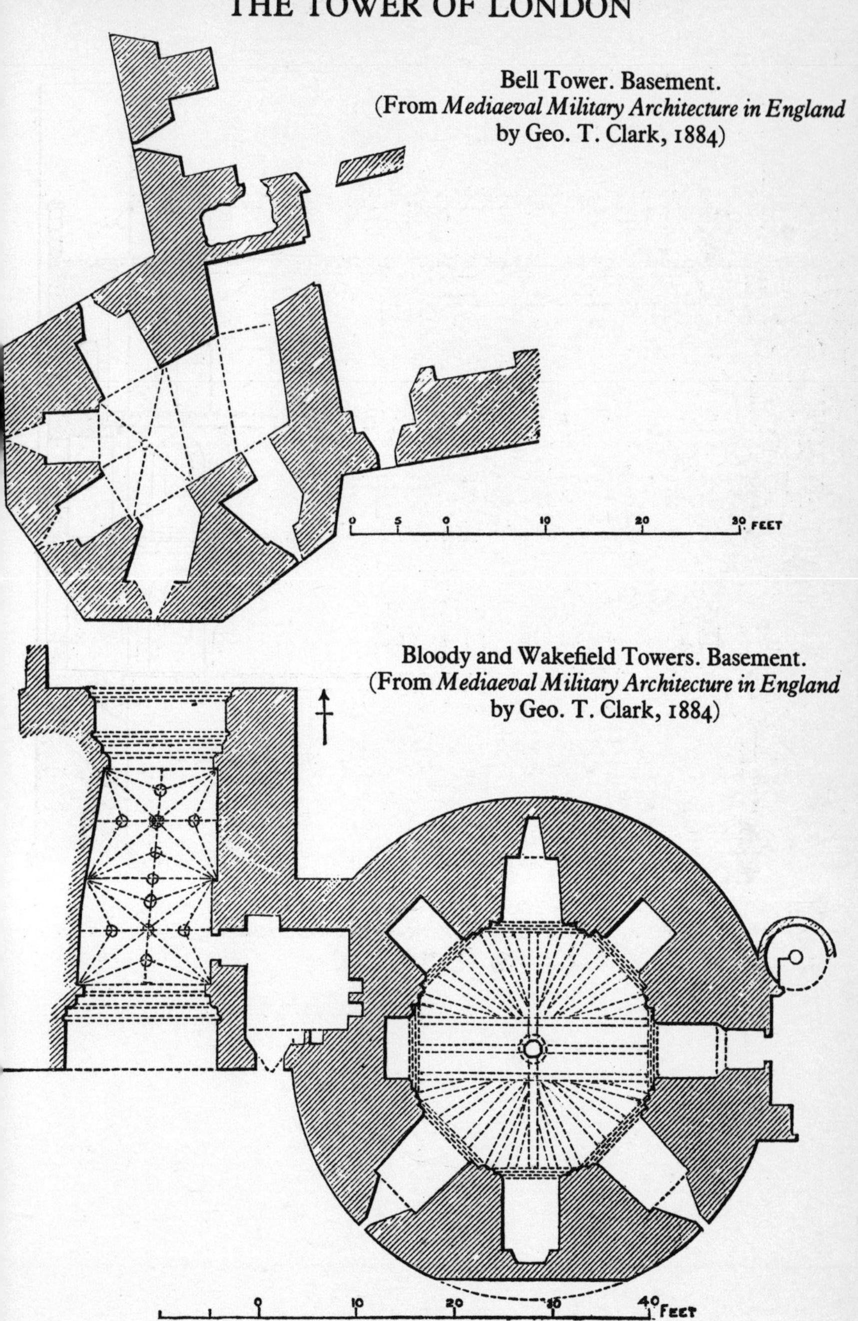

Bell Tower. Basement.
(From *Mediaeval Military Architecture in England*
by Geo. T. Clark, 1884)

Bloody and Wakefield Towers. Basement.
(From *Mediaeval Military Architecture in England*
by Geo. T. Clark, 1884)

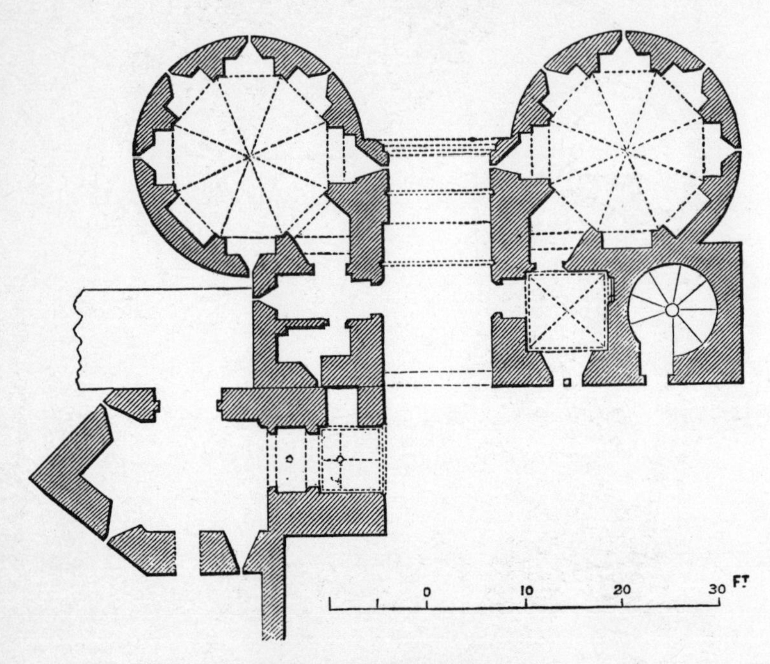

Byward Tower and Postern.
(From *Mediaeval Military Architecture in England*
by Geo. T. Clark, 1884)

Nottingham Castle in the Sixteenth Century.
(From *Mediaeval Military Architecture in England*
by Geo. T. Clark, 1884)

LIST OF ILLUSTRATIONS.

FIG.		PAGE
1	Wooden Ramparts of Roman work, from Trajan's Column	4
2	German Rampart of wood and wicker-work, from the Column of Antoninus	6
3	Wooden Towers on Roman Walls, from Trajan's Column	8
4	Plan of Famars, in Belgium	9
5	Roman Method of constructing the Walls of a Fortification	9
6	Plan of one of the Towers of Carcassonne	11
7	Inside View of the same Tower with its Curtains, from the City	11
8	Outside View of the same Tower	12
9	Bird's-eye View of a Roman Town	14
10	Bird's-eye View of part of a Fortress on a Hill	16
11	Plan of the Roman Walls of Carcassonne	17
12	Section of a Roman Tower, as described by Cæsar	20
13	Attack of a Palisade with a Battering-ram, from a MS. of the Tenth Century	25
14	Ezekiel, with three Battering-rams, from a MS. of the Eleventh Century	26
15	Part of Carcassonne defended by Wood-work when a breach was made	41
16	Plan of Carcassonne as fortified by S. Louis	48
17	Plan of the Castle of Carcassonne	56
18	Bird's-eye View of Carcassonne	57
19	A Curtain-wall with Battlement and Loopholes; and the Wood-work, shewing one mode of attack and defence	61
20	The Lines of Approach	63
21	Attack by the Drawbridge from the Cat	65
22	Plan of one Bay of the Curtain-wall, and two Bastions or Towers, Carcassonne	66
23	Wooden Door of a Bastion	67
24	Plan of Paris in the Thirteenth Century	74

FIG. PAGE

25 Plan of Paris in the Fourteenth Century . . . 75
26 Castle of Coucy 76
27 Plan of the Château-Gaillard and its Environs . . 82
28 Ground-plan of the Château-Gaillard . . . 85
29 Château-Gaillard—Plan of Segments . . . 88
30 Château-Gaillard—View of part of the Wall . . . 89
31 Keep of the Château-Gaillard 91
32 Plan of the Castle of Montargis 95
33 Plan of the Castle of Chauvigny 97
34 Plan of the Keep of Etampes 99
35 Ground-plan of the Keep of Provins . . . 100
36 Plan of the First Story of the Keep of Provins . . 101
37 Plan of Third Story 101
38 Elevation of the Keep of Provins, on the line RS on the Plans 102
39 Section of the Keep of Provins, on the line AB on the Plans 103
40 Plan of the Castle of Coucy 106
41 Ground-plan of the Keep of Coucy 107
42 Plan of the First Story of the Keep of Coucy . . 108
43 Plan of the Second Story of the Keep of Coucy . . 109
44 Plan of the Platform (on the Roof, the Allure behind the Parapet, and of the Battlements) of the Keep of Coucy . 110
45 Section of the Keep of Coucy, on the line O P of the Plans 111
46 Elevation, Section, and Plans of a Tower at Carcassonne . 117
47 A Crenelle with its wooden Hanging Shutter . . 118
48 Plan of part of a Curtain-wall with a Bastion . . 119
49 Plan of one Bay of a Curtain-wall with part of two Bastions 120
50 Plan of a Horn 121
51 Beaks of Loches and of the Gate of St. John at Provins . 122
52 Beaks of the Gates of Jouy at Provins, and of Villeneuve-le-Roi 122
53 Plan of the Town of Aigues-Mortes 123
54 Plan of an Angle of the Fortifications of Carcassonne . 124
55 Plan of a Projecting Angle of the Castle of Falaise . . 124
56 Plan of the Narbonne Gate of the City of Carcassonne . 126
57 Plan of the First Floor of the Narbonne Gate . . 128
58 Elevation of the Narbonne Gate 129
59 Plan of the Upper Story of the Narbonne Gate . . 131
60 The Drawbridge 132
61 Entrance to the Castle of Montargis . . . 133
62 The Tapuca, or Shutter, suspended from above . . 134
63 A Shutter balanced on a Pivot 134

FIG. PAGE

64 Gate of Aubenton, attacked by the Count of Hainault, from a MS. of Froissart 135
65 Plan of the Hoarding 137
66 Sections of part of a Curtain-wall, well defended . . 139
67 Battlements and Machicoulis of a Tower . . . 140
68 Newcastle-on-Tyne, from a MS. of Froissart . . 141
69 Part of the Castle of Pierrefonds 142
70 Part of the Castle of Pierrefonds, *restored* . . . 143
71 Plan of a Square Tower 145
72 Part of the Walls of Avignon, inside . . . 146
73 Ground-plan of one of the Towers of Avignon . . 147
74 Plan of the First Story 147
75 Plan of Upper Story, with the Allure and Battlement . 148
76 Perspective View of the Interior of one of the Towers of Avignon 149
77 Part of the Palace of the Pope at Avignon . . . 150
78 Plan of the Castle of Vincennes 152
79 A Double Cannon with the Wooden Shield or Mantelet . 172
80 A Double Cannon with the Chamber for Powder . . 172
81 A Cannon mounted on a Carriage with a Quadrant . 172
82 An Archer with his Sheaf of Arrows . . . 173
83 An Archer firing downwards 174
84 A Crossbow-man with his Shield on his back, taking aim. From a MS. of Froissart 174
85 A Crossbow-man fitting the handle. From a MS. . . 175
86 The Cranequin, or Handle of the Crossbow . . . 175
87 Plan of the Castle of Bonaguil 178
88 Bird's-eye View of the Castle of Bonaguil . . . 180
89 Embrasure of the Castle of Bonaguil . . . 181
90 Plan of the Walls of the Town of Langres . . . 183
91 Ground-plan of the Great Tower, Langres . . . 184
92 Plan of the First Story 185
93 Section of the Tower, Langres 185
94 Plan of one of the Bastions of Langres, Fifteenth Century 185
95 Section of the Bastion on the line C D of the Plan, fig. 94. 186
96 Section of the Bastion on the line A B of the Plan, fig. 94. 186
97 Plan of a Tower at Perigueux 188
98 View of the Tower at Perigueux 189
99 Fortifications of the Bridge over the Rhine at Schaffhausen 191
100 Plan of the Citadel of Schaffhausen 192
101 Perspective View of one of the Bastions, Schaffhausen . 193

FIG.		PAGE
102	Plan of the first Story of the Bastion, Schaffhausen	194
103	Plan and Section of one of the Embrasures of the Great Hall	195
104	Plan of the Platform of the Tower, Schaffhausen	196
105	Bird's-eye View of Sahaffhausen	198
106	View of a Battery, Schaffhausen	200
107	View of part of the Fortifications of the Town of Orange	201
108	Section of a Parapet at Orange, called a Braie	202
109	View of part of the Fortifications of Sienna	206
110	View of the Parapet of the Curtain-wall, inside	208
111	View of a Parapet shewing the Construction	209
112	Fascines	210
113	Rampart formed of the Trunks of Trees	211
114	Rampart formed of Branches of Trees	211
115	Embrasures formed with Gabions	212
116	View of the Trenches, with Gabions, &c.	213
117	The Mazelle Gate and Barbican at Metz	214
118	View of Barbican, or Boulevard, Metz	216
119	Part of the Fortifications of the Town of Hull	218
120	Fortifications of Lubeck	219
121	Bird's-eye View of the Castle of Milan	220
122	View of the Bridge of Marseilles	223
123	Cavalier on a Bastion at Verona	224
124	Traverses with Gabions	225
125	View of one of the Bastions of Augsburg	230
126	Plan of Bastions at Augsburg	231
127	Ground-plan of the Fortifications of Augsburg	232
128	View of the Fortifications of Frankfort-on-the-Maine	233
129	Plan of one of the Bastions	234
130	View of an Orillon, or Oblong Bastion	235
131	Plan of Orillons	236
132	Plan of one of the Bastions at Troyes	238
133	Plans of Bastions	238
134	View of Bastions attacked	239
135	Bastion isolated, with Inner Rampart	241
136	Bird's-eye View of a Bastion	242
137	Plans of a Ravelin and two Tenailles	244
138	Plan of an Embrasure at Nuremberg	246
139	Section of the same	246
140	View of the Parapet at Nuremberg, with the Hoarding	247
141	Embrasures with Redents	248
142	Plan and View of an Embrasure	248

FIG. PAGE

143 Plan of another Embrasure 248
144 Covered Way, with Crenelles, Loopholes, and Shutters . 249
145 Elevation, Section, and Plan of an Embrasure, with Loopholes for Musketry 250
146 Embrasure of the Laufer Gate at Nuremberg . . 251
147 Bird's-eye View of part of the City of Metz . . 258
148 Plan of Vaulted Gallery 260
149, 150 Plan and Section of a Bastion according to De Ville . 262
151 Section of Ditch with false Braie, according to De Ville . 264

Military Architecture

ESSAY

ON THE

MILITARY ARCHITECTURE

OF THE

MIDDLE AGES.

TO write a general history of the art of fortification, from the days of antiquity to the present time, is one of the fine subjects lying open to the researches of archæologists, and one which we may reasonably hope to see undertaken; but we must admit that it is a subject, to treat which fully requires much and varied information,—since to the knowledge of the historian should be superadded in him who would undertake it the practice of the arts of architecture and military engineering. It is difficult to form an exact estimate of a forgotten art, when we are unacquainted with that art as it is practised in the present day; and in order that a work, of the nature of that which we wish to see undertaken, should be complete, it ought to be executed by one who is at once versed in the modern art of the defence of strong places, an architect, and an antiquary. The present writer is not a military engineer and scarcely an antiquary: it would, therefore, be in the highest degree presumptuous were he to offer this summary in any other light than as an essay,—a study of one phase of the art of fortification, comprised between the establishment of the feudal power, and the definite adoption of the modern system of fortification as devised to counteract the use of artillery. This essay, perhaps, by lifting the veil which still envelopes one branch of the art of mediæval architecture, may induce some of our young officers of engineers to devote themselves to a study, which could not fail to possess great interest, and which might probably have a useful and a practical result; for there is always something to be gained by informing ourselves of the efforts made by those who have preceded us in the same path, and by following up the progress of human labour, from its first rude essays, to the most remarkable developments of the intelligence and the genius of man. To see how others have conquered before us the difficulties by which they were surrounded, is one means of learning how to conquer those which every day present themselves; and in the art of fortification, where everything is a problem to be solved, where all is calculation and foresight, where we have not only to do battle with the elements and with the hand of time, as in the other branches of architecture, but to protect ourselves against the intelligent and previously-planned destructive agency of man, it is well, we think, to know how in past times some have applied all the abilities of their minds and all the material force at their command to the work of destruction, others to that of preservation.

At the time when the barbarians invaded Gaul, many of the towns still preserved their fortifications of Gallo-Roman origin; those which did not, made haste to erect some, out of the ruins of civil buildings. Those walled enclosures, successively forced and repaired, were long the only defensive works of these cities; and it is probable that they were not built upon any regular or systematic plan, but constructed very variously, according to the nature of the localities and of the materials, or after certain local traditions, the nature of which we cannot at the present day fully understand, as there remain to us only the ruins of these walls, consisting of foundations which have been modified by successive additions.

The Visigoths took possession, in the fifth century, of a great portion of Gaul; their domination extended, under Wallia, from the Narbonaise to the Loire. During eighty-nine years Toulouse remained the capital of this kingdom, and, in the course of that period, the greater number of the towns of Septimania were fortified with great care, and had to stand several sieges. Narbonne, Béziers, Agde, Carcassonne, and Toulouse were surrounded by formidable ramparts, constructed according to the Roman traditions of the Lower Empire, if we may judge at least by the important portions of the early walls which still surround the city of Carcassonne. The Visigoths, allies of Rome, did no more than perpetuate the acts of the Empire, and that with some degree of success. As for the Franks, who had preserved their Germanic customs, their military establishments would naturally be so many fortified camps, surrounded by palisades, ditches, and some embankments of earth. Timber plays an important part in the fortifications of the first centuries of the middle ages. And although the Germanic races who occupied Gaul left the task of erecting churches and monasteries, palaces and civil structures, to the Gallo-Romans, they were bound to preserve their military habits in the presence of the conquered nation. The Romans themselves, when they made war upon territories covered with forest, like Germany and Gaul, frequently erected ramparts of wood; advanced works, as it were, beyond the limits of their camps; as we may see by the bas-relief on Trajan's Column (1). In the time of Cæsar, the Celts, when they found themselves unable to continue their wars, placed their women, their children, and all the most precious of their possessions behind fortifications made of wood, earth, or stone, beyond the reach of their enemy's attack.

"They employ," says Cæsar in his Commentaries, "pieces of wood perfectly straight, lay them on the ground in a direction parallel to each other at a distance apart of two feet, fix them transversely by means of trunks of trees, and fill up the voids

Fig. 1. Wooden Ramparts of Roman work, from Trajan's Column.

with earth. On this first foundation they lay a layer of broken rock in large fragments, and when these are well cemented, they put down a fresh course of timber arranged like the first; taking care that the timbers of these two courses do not come into contact, but rest upon the layer of rock which intervenes. The work is thus proceeded with, until it attains the height required. This kind of construction, by reason of the variety of its materials, composed of stone and wood, and forming a regular wall-surface, is good for the service and defence of fortified places; for the stones which are used therein hinder the wood from burning, and the trees being about forty feet in length, and bound together in the thickness of the wall, can be broken or torn asunder only with the greatest difficulty[a]."

Cæsar renders justice to the industrious manner in which the Gallic tribes of his time established their defences and succeeded in resisting the efforts of their assailants, when he laid siege to the town of Avaricum, (Bourges).

"The Gauls," he says, "opposed all kinds of stratagems to the wonderful constancy of our soldiers: for the industry of that nation imitates perfectly whatever they have once seen done. They turned aside the hooks (*falces murales*) with nooses, and when they had caught hold of them firmly drew them in by means of engines, and undermined the mound the more skilfully for the reason that there are in their territories extensive iron-mines, and consequently every kind of mining operation is known and practised by them. They had furnished, moreover, the whole wall on every side with turrets, and had covered these with hides. Besides, in their frequent sallies by day and night they attempted either to set fire to the mound, or attack our soldiers when engaged in the works; and, moreover, by means of beams spliced together, in proportion as our towers were raised, together with our ramparts, did they raise theirs to the same level[b]."

The Germans constructed, also, ramparts of wood crowned with parapets of osier. The Column of Antonine at Rome furnishes a curious example of this kind of rustic redoubt (2). These works were, however, very probably of hasty construction. We see here the fort attacked by Roman soldiers. The infantry, in order to get close to the rampart, cover themselves with their shields and form what was called *the tortoise* (*testudo*); by resting the tops of their shields against the rampart, they were able to sap its base or set fire to it, safe, comparatively,

---

[a] Cæsar, *De Bello Gall.*, lib. vii. cap. 22.   [b] Ibid.

from the projectiles of the enemy [c]. The besieged are in the act of flinging stones, wheels, swords, torches, and fire-pots upon the tortoise; while Roman soldiers, holding burning brands, appear to await the moment when the tortoise shall have completely reached the rampart, in order to pass under the shields and fire the fort. In their entrenched camps, the Romans, besides some advanced works constructed of timber, frequently erected

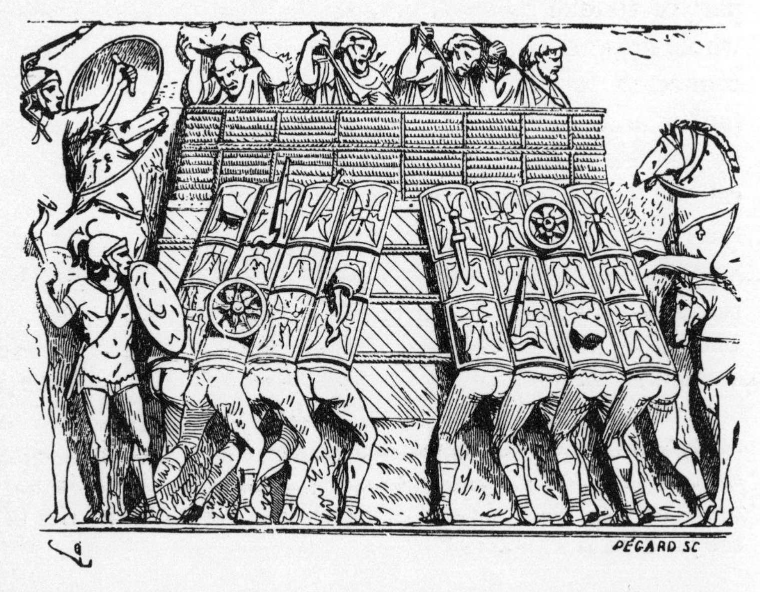

Fig. 2. German Rampart of wood and wicker-work, from the Column of Antoninus.

along their ramparts, at regular intervals, wooden scaffoldings, which served either for placing in position the machines intended to hurl their projectiles, or as watch-towers from which to reconnoitre the approaches of the enemy. The bas-reliefs of Trajan's Column afford numerous examples of this kind of structure (3). These

[c] These shields, formed like a portion of a cylinder, were reserved for this kind of attack.

Roman camps were of two sorts: there were the summer camps, the *castra æstiva*, of a purely temporary nature, which were raised to protect the army when halting in the course of the campaign, and which consisted merely of a shallow ditch and a row of palisades planted along the summit of a slight embankment; and the winter, or stationary camps, *castra hiberna*, *castra stativa*, which were defended by a wide and deep ditch, and by a rampart of sodded earth or of stone flanked by towers; the whole crowned with crenellated parapets or with stakes, connected together by means of transverse pieces of timber or wattles. The use of round and square towers by the Romans in their fixed entrenchments was general, for, as Vegetius says,—

> "The ancients found that the enclosure of a fortified place ought not to be in one continuous line, for the reason that the battering-rams would thus be able too easily to effect a breach; whereas by the use of towers placed sufficiently close to one another in the rampart, their walls presented parts projecting and re-entering. If the enemy wishes to plant his ladders against, or to bring his machines close to, a wall thus constructed, he can be seen in front, in flank, and almost in the rear; he is almost hemmed in by the fire from the batteries of the place he is attacking."

From the very earliest antiquity the usefulness of towers had been recognised for the purpose of taking the besiegers in flank when they attacked the curtains.

The fixed camps of the Romans were generally quadrangular, with four gates pierced, one in the centre of each of the fronts; the principal gate was called the *prætorian*, because it opened in front of the *prætorium*, or residence of the general-in-chief; the opposite one was called the *decumana;* the two lateral gates were known as *principalis dextra* and *principalis sinistra.* Outworks, called *antemuralia*, *procastria*, defended those gates[d]. The officers and soldiers were lodged in huts built of clay,

Fig. 3. Wooden Towers on Roman Walls, from Trajan's Column.

brick, or wood, and thatched or tiled over. The towers were provided with machines for hurling darts or stones. The local position very often modified this quadrangular arrangement, for, as Vitruvius justly observes, in reference to machines of war (cap. xxii.),—"As for the means which a besieged force may employ in their defence, this cannot be set in writing."

The military station of Famars, in Belgium (*Fanum Martis*), given in the "History of Architecture in Belgium," and the plan of which we here produce (4), shews an enclosure, of which the arrangement is not in accordance with the ordinary plans of Roman camps: it is true, this fortification cannot be referred to an earlier date than the third century[e]. As for the mode adopted by the Romans

---

[d] Godesc. Stewechii, *Conject. ad Sexti Jul. Frontini lib. Stragem.* Lugd. Batav., 1592, 12mo., p. 465.

[e] See *Hist. de l'architect. en Belgique*, par A. G. B. Schayes, t. i. p. 203. (Bruxelles.)

in the construction of their fortifications for cities, it consisted in two strong walls of masonry, separated by an interval of twenty feet: the space between was filled with the earth from the ditches, and loose rock well rammed, forming at top a parapet walk, slightly inclined towards the town to allow the water to pass off: the outer of these two walls, which was raised above the parapet-walk, was massive and crenellated; the inner one was very slightly elevated above the ground level of the place inside, so as to render the ramparts easy of access, by means of flights of steps and inclined ways (5).[1]

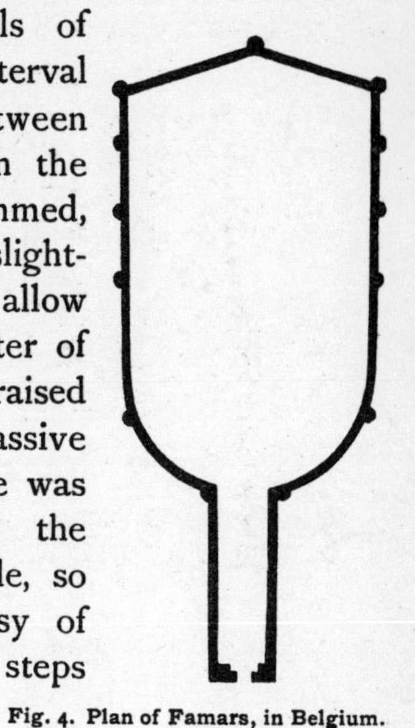

Fig. 4. Plan of Famars, in Belgium.

The *Chateau Narbonnais* at Toulouse, which plays so

Fig. 5. Roman Method of constructing the Walls of a Fortification.

important a part in the history of that city from the time of the domination of the Visigoths to the fourteenth

[1] Vegetius, lib. iv. cap. 3. tit. *Quemadmodum muris terra jungatur egesta.*

century, appears to have been constructed according to the classical model: it was composed of—

"Two massive towers, one at the south, the other at the north, built of baked clay and flint, with lime; the whole enclosed by great stones without mortar, but cramped together by means of iron plates run with lead. The castle stood above the ground level more than thirty fathom (*brasses*)[g], having towards the south two successive gates and two vaults of masonry reaching to the summit of the building; there were also two other successive gates on the north side and on the *Place du Salin*. By the latter of these gates you formerly entered the city, the ground of which has been since raised more than twelve feet . . . A square tower was to be seen between these two towers, or defensive platforms; for they were embanked and filled with earth, according to Guillaume de Puilaurnes, since it appears that Simon de Montfort had all the earth removed which then filled them to their roofs[h]."

The Visigoth fortification of the city of Carcassonne, which is still preserved, offers an analogous arrangement, recalling those described by Vegetius. The level of the town is much more elevated than the ground outside, and almost as high as the parapet walks. The curtain walls, of great thickness, are composed of two faces of small cubical masonry alternating with courses of brick; the middle portion being filled, not with earth, but with rubble run with lime. The towers were raised above these curtains, and their communication with the latter might be cut off, so as to make of each tower a small independent fort; externally, these towers are cylindrical, and, on the side of the town, square: they rest also, towards the country, upon a cubical base or foundation. We subjoin (6) the plan of one of these towers with the curtains adjoining.

---

[g] The *brasse*, or Fr. fathom, measured 5 *pieds du Roy*.

[h] *Annales de la ville de Toulouse*. Paris, 1771, t. i. p. 436.

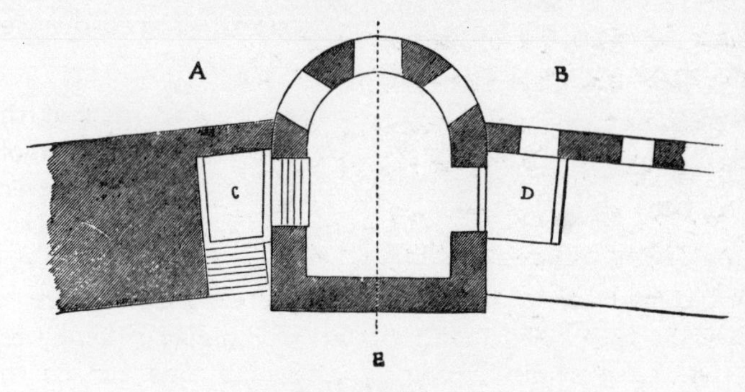

Fig. 6. Plan of one of the Towers of Carcassonne.

A. Ground-plan. B. Plan of first story. C & D. Pits beneath drawbridges.

Fig. 7. Inside View of the same Tower with its Curtains, from the City.

A is the plan on the ground level; B, the plan of the first story at the level of the parapet. We see, at C and D, the two excavations formed in front of the gates of the tower to intercept, when the drawbridges were raised, all communication between the town, or the parapet walk and the several stories of the tower. From the first story, access was had to the upper crenellated, or battlemented, portion of the tower by a ladder of wood placed interiorly against the side of the flat wall. The external ground-level was much lower than that of the tower, and also beneath the ground-level of the town, from which it

Fig. 8. Outside View of the same Tower.

was reached by a descending flight of from ten to fifteen steps. Fig. 7 shews the tower and its two curtains on the side of the town; the bridges of communication are supposed to have been removed. The battlemented portion at the top is covered with a roof, and open on the side of the town, in order to permit the defenders of the tower to see what was going on therein, and also to allow of their hoisting up stones and other projectiles by means of a rope and pulley[i]. Fig. 8 shews the same tower on the side towards the country; we have added a postern[k], the sill of which is sufficiently raised above the ground to necessitate the use of a scaling or step ladder, to obtain ingress. The postern is defended, as was customary, by a palisade or barrier, each gate or postern being provided with a work of this kind.

In conformity with the traditions of the Roman fixed camp, the fortifications of the towns of the middle ages enclosed a castle, or at the least a fort, which commanded the walls; the castle itself contained a detached defence stronger than all the others, which took the name of donjon. Frequently the towns of the middle ages were protected by several fortified walls, one within the other; or there was the city proper, which, placed upon the point of greatest elevation, was surrounded by strong walls, and around it faubourgs (or suburbs) defended by towers and curtains, or by simple works of earth or timber, with ditches. When the Romans founded a city, they took care, as far as was possible, to choose some site sloping towards a river. When the inclination of the ground was terminated by another embankment, sloping in the opposite direction, at some distance from

---

[i] These towers were partially damaged at the beginning of the twelfth century, after the taking of Carcassonne by the army of Saint Louis. At several points, however, may be seen traces of these interruptions between the curtains and the gates of the tower.

[k] This postern exists at the side of one of the towers and is protected by its flank.

the course of the river, the site fulfilled all the conditions to be desired. We give (9), in order to make ourselves better understood, a bird's-eye view of the site of a Roman city, according to the above data. A was

Fig. 9. Bird's-eye View of a Roman Town.

A. The town. B. The escarpment. CC. The walls.
D. The castle. EE. The watch-towers.

the city, with its walls bounded on one side by the river; frequently a bridge, defended by advanced works, communicated with the opposite bank. At B was the escarpment, which rendered access to the town difficult at the point where an enemy's army would naturally attempt to invest it; D, the castle commanding the whole system of defence, and serving as a refuge for the garrison in case the city should fall into the enemy's hands. The weakest points were thus the two fronts, CC, and therefore it was here that the walls were high, well flanked by towers, and protected by wide and deep ditches, sometimes also by palisades, more especially in advance of the gates. Neither was the position of the besiegers, when facing either of these two fronts, very good; for a sally which would take them in flank might, were the garrison at all brave and numerous, drive them back into the river. With a view to reconnoitring the operations of the besieging army, there were erected, at the angles EE, towers of great elevation, which allowed those in the town to watch the banks of the river both up and down to a great distance, and also the two fronts CC. It is according to this arrangement that the cities of Autun, Cahors, Auxerre, Poitiers, Bordeaux, Langres, &c., were fortified in the Roman times. When a bridge connected, in front of the walls, the opposite sides of the river, then the bridge was defended by a *tête-de-pont*, G, on the side over against the town. These *têtes-de-pont* assumed more or less importance in different places; they took in whole suburbs, or were merely fortresses, or simple barbicans. Stockades, with towers face to face, built on the two banks of the river above the bridge, permitted the townspeople to bar the passage and intercept the navigation by throwing from one tower to the other either chains, or pieces of wood attached end to end by iron rings. If, as was the case with Rome herself, in the neighbourhood of a river were situated a series of hills, care was taken not to surround these hills, but to carry the walls of defence across their summits; fortifying strongly at the same time the intervals, which, being commanded by the front, on both sides, could not be attacked without great risk. For this purpose, also, between the hills the line of the walls was nearly always inflected and concave in such a way as to flank the valleys, as is shewn in the bird's-eye view (10)[1]. But

[1] See the plan of Rome.

if the city stood in the plain (in which case it was generally of secondary importance), advantage was taken of

Fig. 10. Bird's-eye View of part of a Fortress on a Hill.

every rise in the ground; the sinuosities being carefully followed, so as to prevent the besiegers from establishing themselves on a level with the foot of the walls, as may be seen at Langres and Carcassonne,—we append (11) the Visigoth enceinte of the latter town—we might almost say the Roman one, inasmuch as some of the towers are built on Roman foundations. In the cities of antiquity, as well as in the greater number of those erected in the middle ages, and in those of our own day, the castle (*château*, *castellum; capdhol*, *capitol* in *langue d'oc*) was built, not only on the point of greatest elevation, but also contiguous on one of its sides to the city wall, in order to secure to the garrison the means of receiving succour from without if the city were taken. The entrances into the castle were protected by outworks, which extended a considerable distance into the country, so as to leave between the first barriers and the walls of the castle an open space, or *place d'armes*, which would allow of the encampment of a body of troops beyond the fixed lines of fortification, to sustain the shock of the first attacks. These advanced intrenchments were generally thrown up in a semicircular line and

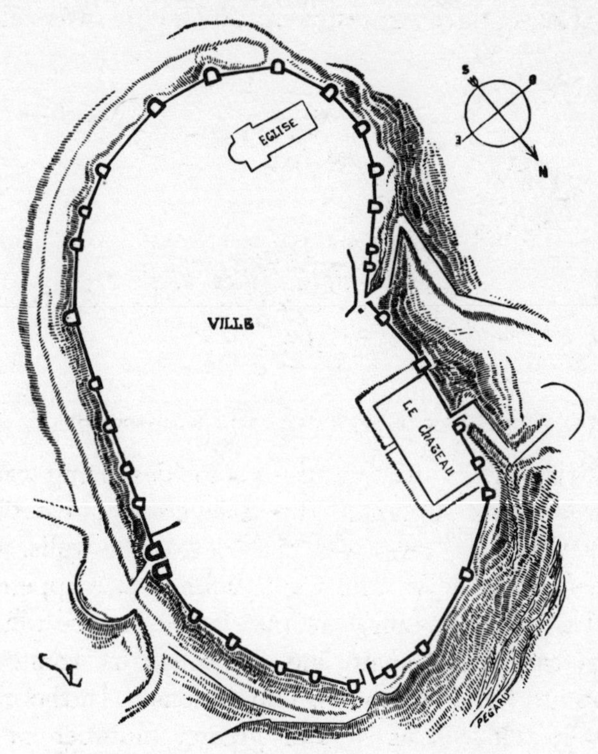

Fig. 11. Plan of the Roman Walls of Carcassonne.

composed of ditches and palisades; and the gates were placed laterally, so as to oblige the enemy who endeavoured to force them to present himself in flank before the walls of the place.

As from the fourth to the tenth century the defensive system of Roman fortification had undergone but little modification, the means of attack had necessarily lost much of their power; the mechanical arts played an important part in the sieges of fortified places, and practical mechanics were not likely to be developed, or indeed to maintain the level to which the Romans had raised them, under the domination of barbarian conquerors.

The Romans were very skilful in the art of attacking strong places, and they displayed under those circumstances a vastness of resources of which we can hardly form an adequate idea. Their military organisation was, moreover, in the highest degree favourable to the war of sieges: all their troops could be converted into pioneers, labourers, miners, carpenters, masons, &c., and an army *en masse* laboured at the approaches, the earth-works, the walls of contravallation, at the same time that they attacked the enemy and defended themselves. Herein lies the explanation of the fact that Roman armies, comparatively not numerous, brought to a successful issue sieges in the course of which they had been obliged to construct gigantic works. When the Roman lieutenant, C. Trebonius, was left by Cæsar at the siege of Marseilles, the Romans had to erect considerable works in order to reduce the city, which was strong and well provided with means of defence. One of their works of approach is of great importance: we give here the passage of Cæsar's Memoirs which describes it, endeavouring in our translation to render it as intelligible as possible:—

"The legionaries, who directed the right of the work, considered that a tower of brick, erected at the base of the wall (of the town), might be of considerable assistance to them against the frequent sallies of the enemy, if they succeeded in making it into a fort or bastille. That which they had first made was small and low; it served them, however, as a place of retreat. In it they defended themselves against superior forces, or they issued from it to repulse and pursue the enemy. This work was thirty feet long on each of its sides, and the thickness of the walls was five feet; it was soon discovered (for experience is a great master) that a great advantage might, by means of some additions to the original plan, be taken of this structure, if it were given the elevation of a tower.

"When this fort had been carried up to the height of one

story, they (the Romans) laid down a floor composed of joists, the ends of which were covered by the external face of the masonry, in order that the fire thrown by the enemy could not fasten upon any projecting portion of the wood-work. Above this floor they raised the brick walls as much as they were allowed by the parapets and mantelets by which they were screened; then, at a short distance from the coping of the walls, they laid two diagonal beams to carry the framework intended to form the roof of the tower. Upon these two beams they set transverse joists on a radiating plan, the extremities whereof were allowed to overhang somewhat the external face of the tower wall, in order to suspend from them, outside, guards which would shield the workmen engaged on the construction of the wall. They covered this framework with bricks and clay to render it fire-proof, and stretched a rude kind of temporary covering over it, lest the roof should be beaten in by the projectiles thrown by the engines, or the bricks broken by the stones from the catapults. They then made three mats with cables such as are used for holding the anchors of vessels, of the length of each of the sides of the tower and the height of four feet, which they fastened to the external extremities of the beams (of the roof) along the walls, on the three sides facing the enemy. The soldiers had often had proof, upon other occasions, that this kind of guard was the only one which formed an impenetrable barrier against the arrows and projectiles hurled from the engines. A portion of the tower being complete and placed beyond the reach of assault, they transferred the mantelets they had used to other parts of the attacking works. Then supporting themselves upon the first floor, they began to hoist up the whole roof, of a piece, and raised it to a height sufficient to allow the cable-screens still to cover the labourers. Hidden behind these guards, they went on building the walls, which were of brick, then raised the roof a little more, and thus secured for themselves the necessary space for raising their structure by degrees. When they had reached the height of another story, they laid another floor of joists, the bearings of which were always concealed by the external masonry; and from thence they continued to raise the roof with its hanging cable-work. Thus it was that, without running any risk, without exposing themselves to be wounded, they successively raised the work six stories. Loop-holes were left, in proper positions, to receive machines of war.

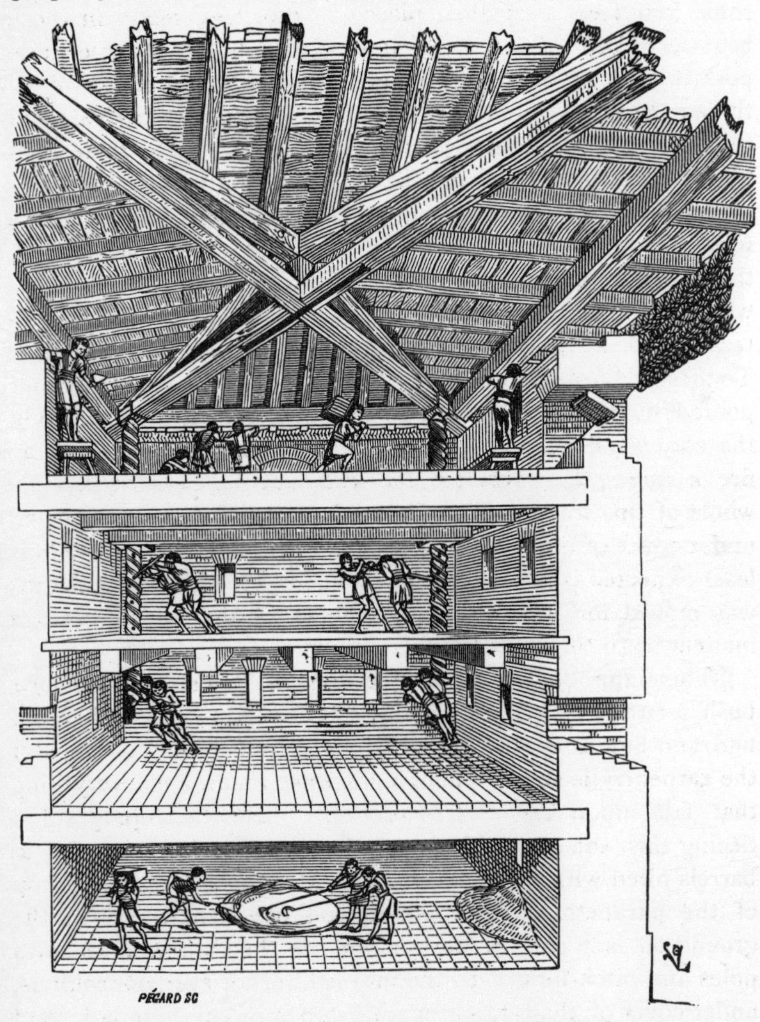

Fig. 12. Section of a Roman Tower, as described by Cæsar.

"When they had made sure that from this tower they could defend the works which adjoined it, they began to erect a *rat* (*musculus*)[m], sixty feet long, with beams two feet square, which

[m] Isidorus, libro duodevigesimo Etymologiarum, capite de Ariete: *Musculus*,

from the ground-floor of their tower would lead them to that of the enemy and to the walls. To this end they first laid down on the ground two beams of equal length, at a distance apart of some four feet; they then placed in mortices, made in those beams, upright posts five feet in height. They connected these posts by rafters joined in the form of a pediment of low pitch, thereon to place the purlins intended to support the roofing of the rat. Over these they placed purlins of two feet square, connected by means of pins and bands of iron. Upon these purlins were nailed laths of four fingers square, to support the bricks (or tiles) which formed the roof. The timber frame-work being thus constructed, and the lower beams carried on traverses, the whole was lined externally with bricks and moist clay, to protect it against the fire which would be launched from the walls. To these bricks were attached hides, in order to hinder the water poured into channels by the besieged from wetting and detaching the clay; and in order that the hides might not be injured by fire or stones, they were covered with mattresses of wool. The whole of this work was constructed at the base of the tower, under cover of mantelets; and all at once, when the Marseillais least expected it, by means of rollers employed in ships, the rat was moved forward against the tower of the city, in such a manner as to touch its base.

"Then the besieged, affrighted by this rapid manœuvre, push forward by means of levers the hugest stones they can find, and hurl them from the top of the wall upon the rat. But the carpentry is strong enough to resist them, and everything that falls upon the roof is carried off by its sloping sides. Seeing this, the besieged change their plan, and setting fire to barrels filled with pitch and tar, precipitate them from the top of the parapets. These barrels roll down, and fall to the ground on each side of the rat, whence they are removed with poles and pitch-forks. In the meanwhile, however, our soldiers, under cover of the rat, loosen the stones of the foundations of the enemy's tower. The rat is likewise defended by arrows shot from the upper works of our brick tower: the besieged

---

inquit, *cuniculo similis sit, quo murus perfoditur: ex quo et appellatur, quasi marusculus.* (Godesc. Stewec., comm. ad lib. iv. Veget., 1492.)

are driven from the parapets of their towers and curtains; no time is left them to shew themselves thereat, or for defence. Already a great quantity of the stones of the basement are removed, when all at once a portion of the tower falls down[n]."

In order to render this passage intelligible, we give (fig. 12) a perspective section of the tower (or bastille) here described by Cæsar, supposed to be taken at the moment when the Roman soldiers are engaged in raising it, under shelter of the moveable roof. This latter is lifted at the four angles by means of large wooden screws, the threads of which are made to work successively in large nuts in two pieces and supported by the first lateral beams of each story and at the angles of the tower. In this way, those screws are endless, for when they leave the nuts of one of the lower stories, they have already entered the nut of the last floor laid: holes pierced in the body of these screws allow six men, at least, to turn them by means of bars, as in a capstan. According as the roof is raised, masons prop it at several points and adjust it to a true level. From the extremities of the beams of the roof are suspended the cable mats to protect the workmen. As to the rat, or gallery intended to enable the pioneers to sap the base of the walls of the besieged under cover, its description is sufficiently clear and detailed in the text not to need a commentary.

If the sieges undertaken by the Romans denote amongst this people great experience and habits of method carefully followed out, the military art in a high state of development, the use of means which were then irresistible, a perfect order in all operations, the same cannot be said of the barbarians who invaded the West; and if the German tribes of the East and North were

[n] Cæsar, *De Bello Civ.*, lib. ii. cap. 8—11.

able to penetrate easily into Gaul, the reason is rather to be looked for in the weakness of the defensive fortifications there than in any skilfulness of their modes of attack, for the vestiges of Roman warfare were hardly known to the barbarians. The few documents which remain to us having reference to the sieges undertaken by the tribes who invaded Gaul, exhibit a notable want of experience on the part of the assailants.

The attack calls for greater order, more regularity, than the defence. The German tribes may have had some idea of defensive fortification, but it was difficult for them to keep irregular and ill-disciplined armies together before a town which held out for any length of time; whenever a siege was protracted, the assailant was almost certain to see his troops breaking up, to go and pillage the country. The military organization of the German nations, moreover, did not favour a war of sieges. As each chief preserved a kind of independence, it was not possible to compel an army composed of such various elements to execute the manual labour to which the Roman armies were habituated. The German soldier would have disdained to take in hand the spade and the shovel to make a trench or to throw up an embankment; and it is a matter of certainty that, if the Gallo-Roman cities had been well provided with materials of war, and well defended, the efforts of the barbarians would not have availed against their walls, since, in view of the offensive means at the disposal of their troops, the traditions of the Roman system of defence were superior to the attack. But, after the first invasions, the Gallo-Romans perceived the necessity of defending themselves, and of fortifying their towns, dismantled in the course of a long peace; on the other hand, the barbarian troops had acquired a greater amount of experience, and were not long before they put into practice, with less order indeed, but with a more furious energy and a greater sacrifice of life, the greater number of the means of attack which had been practised by the Romans. Once masters of the soil, the new conquerors put forth their warlike genius in improving the defence and attack of cities; constantly engaged in internecine war, occasions were never wanting of applying the remains of the Roman military art; and the ambition of the chiefs of the Franks, down to the time of Charlemagne, was ever to conquer the ancient predominance of Rome, to lean for support upon the civilization in the midst of which they were thrown, and to resuscitate it for their own advantage.

All the sieges undertaken during the Merovingian and Carlovingian periods are rude imitations of the sieges made by the Romans. When a place was about to be invested, two lines of ramparts of either wood or stone, and protected by ditches, were first of all established; one on the side of the city, in order to afford a protection against the sorties of the besieged and to deprive them of all means of communication with the outside,—this is the line of contravallation; the other on the side of the champaign, or open country, in order to provide against succour reaching the place from without, which is the line of circumvallation. In imitation of the Roman armies, the towers, which formed part of the ramparts attacked, were opposed by other towers of wood, moveable and of a greater height than the former, which commanded the ramparts of the besieged, and which, by means of flying bridges, allowed of numerous assailants being thrown upon the walls. The moveable towers had this advantage, namely, that they could be placed opposite the weak points of the defence, against curtains which had but narrow parapet-ways (*chemins de ronde*), and which, consequently, were able to oppose no more than a single line of soldiers to a deep column of attack; the latter being precipitated, moreover, from a higher point upon the walls. The art of the miner and all the engines constructed for battering walls were greatly improved: and thenceforward the attack overpowered the defence. Of the machines of war of the Romans, the armies of the first centuries of the Middle Ages had preserved the battering-ram (*mouton* in *langue-d'oil*, *bosson* in *langue-d'oc*). This fact has been sometimes doubted; but we have proofs of the use, during the tenth, eleventh, twelfth, fourteenth, fifteenth, and even sixteenth centuries, of this engine for battering walls. We append copies of vignettes taken from manu-

Fig. 13. Attack of a Palisade with a Battering-ram, from a MS. of the Tenth Century.

scripts in the Imperial Library of Paris, which must dispel all doubt as to the use of the battering-ram. The first of these (fig. 13) represents an attack on palisades, or lists, surrounding a fortification of stone [o]. The battering-ram can be plainly distinguished carried upon two wheels and impelled by three men who cover themselves with their shields; a fourth assailant holds a cross-bow.

Fig. 14. Ezekiel, with three Battering-rams, from a MS. of the Eleventh Century.

The second (fig. 14) represents one of the visions of Ezekiel [p]; three battering-rams, on wheels, surround the prophet [q]. In the siege of the castle of Beaucaire by the

[o] Haimonis, *Comment. in Ezech.*, Bibl. Imp., manusc. of the tenth century, F. de Saint-Germain, Latin, 303.

[p] Bible, No. 6. vol. iii. Bibl. Imp., ancient F. Lat. MS. of ninth or tenth century.

[q] Ezekiel iv. 2, 3. Ezekiel is shewn as holding the plate of iron, and the battering-rams surround him.

inhabitants of that town, the *bosson* is employed (see further on, where the use of this engine is alluded to). And finally, in the Chronicles of Froissart, and, later still, at the siege of Pavia under Francis I., mention is made of the battering-ram. But after the first crusades, the engineers of the Western countries who had accompanied the armies to the East brought back with them to France, Italy, England, and Germany some improvements in the art of fortification. The feudal system, already organized, soon put those new methods into practice, with never-ceasing amelioration, owing to the state of permanent war in which it existed. From the close of the twelfth century until towards the middle of the fourteenth the defence continued stronger than the attack; nor did this state of things undergo any change until gunpowder came into use with artillery. From that moment the attack has never ceased to be superior to the defence.

Down to the twelfth century it does not appear that towns were defended otherwise than by fortified walls with flanking towers, or by simple palisades with a ditch, having wooden towers, or bastilles, at intervals; which was the Roman method: but at this time the land was covered with castles, and experience had proved that a castle could defend itself better than a city. In fact, one of the most admitted principles of fortification at the present day consists in opposing the greatest possible front to the enemy; because the greater the front, the greater the envelope which it requires, and the longer and more considerable, therefore, must be the labour of the besiegers. But when it was necessary to have the battering engines close to the walls; when, to destroy the works of the besieged, only the sap, the ram, the mine, or engines of inconsiderable range of projection were employed; when the assault could only be made by means of the wooden towers already described, or by scaling the walls, or through breaches ill-made and difficult of access, the more the garrison was concentrated, the more strength it possessed. For the besieging army, however numerous it might be, when once obliged to come to close quarters, could only have at any given point a force equal, at the utmost, to that of the besieged. On the other hand, walls of great extent, which could be attacked suddenly by a numerous army on several points at the same instant of time, divided the forces of the besieged; and required an army at least equal to the investing force to man the walls properly, and to repel attacks which frequently only became known at the moment of their execution.

In order to do away with the inconvenience arising from having great fronts to fortify, towards the close of the twelfth century the idea was started of establishing, in advance of the main walls with their flanking towers, isolated fortresses or detached forts, intended to keep the assailant at a distance from the body of the place, and to force him to draw out his lines in contravallation to such an extent that an immense army would be required to guard them. With the artillery of modern times, the *con*verging fire of the besieging army gives it a superiority over the *di*verging fire of the besieged; but before the invention of cannon the attack could only be made within a very short distance of the walls, and always in a direction *perpendicular to the front attacked.* The besieged had, therefore, an advantage in opposing to the assailant isolated points, not commanding one another, but well defended; the forces of the enemy were thereby scattered, as he was thus obliged to undertake several simultaneous attacks, upon points chosen by the besieged, and of course well furnished by them with the means of resistance. If the assailant left in his rear those isolated strongholds, in order to attack the fronts of the city, he was open to be taken in the rear by the garrisons of the detached forts just at the moment of delivering the assault, and his position was therefore bad. Sometimes, in order to avoid having to lay siege regularly to each of these forts, the besiegers, if they had a numerous army, erected *bastilles*, or towers of stone laid without mortar, or of timber or earth, such as the Romans were in the habit of erecting, established lines of contravallation around the isolated fortresses, and having thus hemmed in their garrisons, attacked the main body of the place.

All the preliminary operations of sieges were long and uncertain; large supplies of timber and projectiles were required; and it frequently happened that the works of contravallation, the moveable towers, the fixed bastilles of wood and the engines, had hardly been completed, when a vigorous sortie of the besieged or a night attack destroyed by fire and steel the labour of many months. To prevent these disasters, the besieged formed their lines of contravallation with double rows of strong palisades of timber, one behind the other, at the distance of a pike-length, (three to four yards); then, excavating a ditch along the front, they made use of the earth so obtained to fill in the space between the palings; they covered their machines and their wooden towers, moveable and stationary, with ox and horse hides, raw or boiled, or with a kind of thick woollen stuff, so as to render them proof against incendiary projectiles. It often happened that the parts played by the hostile forces were reversed, and that the assailants, driven back by the sorties of the garrisons and forced to take refuge in their camp, became besieged in their turn. In all ages, the works of approach in sieges have been long and beset with difficulties; but in the days of which we write, much more than at the present time, it was the custom of the besieged to make frequent sorties from their walls, either to skirmish at the barriers and prevent the enemy from establishing fixed works, or to destroy the works already executed by the assailants.

Armies were carelessly guarded, as always occurs with irregular and ill-disciplined troops; they trusted to the palisades for keeping out the enemy, and every one relying upon his neighbour for guarding the works, it frequently happened that a hundred or so of men-at-arms, issuing silently from their gates at dead of night, penetrated without meeting a sentinel to the very heart of the encamped army, set fire to their machines of war, and cutting the tent-ropes to increase the disorder, were able to retire before the bulk of the army could get to their legs. In the chronicles of the twelfth, thirteenth, and fourteenth centuries, these surprises are of daily occurrence, nor were the armies a whit more careful on the morrow. It was generally also during the night that, by means of incendiary projectiles, they endeavoured on either side to set fire to the timber works used in the siege.

The Orientals possessed projectiles of this nature which struck great terror into the armies of the West, a fact which would lead us to suppose that they were unacquainted with their composition, at least during the crusades of the twelfth and thirteenth centuries; and they had powerful machines [r] which differed from those

[r] "One evening it happened that the Turks brought an engine, which they called the *pierrière*, a terrible engine of destruction; and they set it up over against the *chaz-chateilz* (a wooden tower in advance of the walls; *vide* Du Cange,

of the Occidentals, as these latter when they adopted them gave them names which indicate their origin—as *Turkish engines, Turkish pierrières.*

We cannot doubt that the crusades, during which so many memorable sieges were effected, improved the means of attack, and that consequently important modifications were introduced into the defence of fortified places. Down to the thirteenth century fortification relies chiefly upon its passive force, on the mass and the position of its walls. It sufficed to enclose a feeble garrison in towers and behind walls of great height and thickness, to enable them to hold out a long time against assailants whose means of attack were weak. The Norman castles, which were erected in such numbers by those new conquerors of the north-west of France and in England, presented masses of building which defied all attempts at escalade because of their height, and which were almost beyond the reach of the sap. The builders, moreover, were always careful to plant, as far as it was possible, these castles upon elevated spots, on some table-land or high-up level of rock, or even on artificial hillocks; to surround them with deep ditches, so as to render it impossible to undermine them; and as a refuge, in the event of surprise or treason, the outer enceinte of the castle contained always an isolated donjon or keep, com-

s. v. *belfredus*, vol. vii. p. 345), by which Messire Gaultier Curel and I kept watch during the night. By the which engine they threw on us the fire *gregeois à planté* (Greek fire), which was the most horrible thing that ever I saw in my life. . . . The manner of the *Greek fire* was in this wise, that in front it was of the bigness of a tun, and the tail of it stretched behind to the length of about half a yard (*aussi grant comme un grant glaive*). It made such a noise in its coming as if it were a thunderbolt falling from heaven, and seemed to me like a great dragon flying in the air, and shewed so great a light, that in our lines it was as light as day, so great a flame of fire was there. Thrice during that night did they cast the said fire from the said *pierrière*, and four times from the catapult of the tower."—(*Joinville, Hist. de Saint Louys.*)

manding the whole of the works, itself frequently surrounded by a moat and a wall (*chemise*), and which from its position generally close to the outside, and the great height of its walls, would enable a few men to hold in check a large body of assailants, or to escape if the place were no longer tenable.

But after the first crusades, and when the feudal system had placed in the hands of some of the nobles a power almost equal to that of the king, it became necessary to discard the system of passive fortification, indebted to its mass only for its defensive power, and adopt a system of fortification which would give to the defence an activity equal to that of the attack, and require at the same time more numerous garrisons. It no longer sufficed (and the terrible Simon de Montfort had proved the fact) to be in possession of massive walls, of castles built upon steep rocks, from the summit of which an assailant without active means of attack might be despised: it was necessary to defend those walls and those towers, and to furnish them with numerous troops, with engines and projectiles; it was necessary to multiply the means of inflicting injury on the besieger, to render all his efforts unavailing, by effecting combinations which he could not foresee, and, above all, to place the garrison beyond the reach of surprises or *coups-de-main*: for it not unfrequently happened that a place of great strength and well furnished with all the munitions of war fell beneath the sudden attack of a small troop of daring soldiers, who, passing over the bodies of the guards at the barriers, seized on the gates, and in this way secured for the main body of the army an entrance into the town.

Towards the end of the twelfth century, and during the first half of the thirteenth, the means of attack and defence, as we have said, were much improved, and especially by their being more methodically carried out. We see, then, for the first time in armies and fortified places, engineers (*ingegneors*) specially intrusted with the construction of the engines intended for attack and defence. Amongst these engines there were some which were at the same time defensive and offensive, that is to say, constructed so as to protect the pioneers and batter the wall; others were offensive merely. When escalade (the first means of attack almost always employed) was not successful, and the gates were too strongly armed to be forced, then it became necessary to undertake a formal siege; it was then that the besiegers erected towers of wood, moving on rollers (*baffraiz*), which they endeavoured to construct loftier than the walls of the town or place besieged, and a kind of moveable platform or gangway called *chat*, *gat*, or *gate*, the Roman *musculus* which Cæsar describes at the siege of Marseilles, formed of wood and covered with planks, iron, and hides, which was pushed to the foot of the walls, and which afforded a covering to the assailants when they wanted either to employ the *mouton* or *bosson* (the battering-ram of the ancients), or to undermine the towers or curtains by the use of pickaxes, or, lastly, to carry forward earth or fascines to fill up the moat.

In the poem of the crusade against the Albigenses, Simon de Montfort frequently uses the *gate*, which appears not only intended to allow the besiegers to sap the foot of the walls under cover, but also to play the part of the moving tower, in raising a body of troops to the level of the parapet:—

"The Count de Montfort commands: . . . Advance ye now the *gate* and ye will take Toulouse. . . . And they (the French) push forward the *gate* with shouts and shrill cries; over the space betwixt the wall (of the town) and the castle it advances with short leaps, like the sparrow-hawk when it hunts down the small birds. Straight forward comes the stone launched by the catapult (*trébuchet*), and strikes it so fierce a blow on its topmost plank, that it breaks open and tears asunder its leathern covering. . . . If you turn the *gate*, cry the barons (to the Count de Montfort), you will save it from these strokes. Par Dieu, says the Count, and we shall try that ere very long. And when the *gate* turns, it goes on again with short and broken leaps. The catapult takes aim, makes ready its charge, and deals it so rude a stroke the second time, that the iron and steel, the beams and bars, are cut and broken."

And further on :—

"The Count de Montfort has gathered together his knights, the bravest and best men of the siege; he has furnished it (the *gate*) with good defences covered with iron on the face, and he has put therein his companies of knights, well covered by their armour, and with their visors down: so they push the *gate* vigorously and quick. But the men of the town are well experienced; they have made ready their catapults, and placed in the slings fine pieces of cut rock, which, when the cords are loosed, fly impetuous, and strike the *gate* on the front and flanks so truly, that from doors and floor, from roof and sides, the splintered timber flies on all sides, and that, of those who drive it forward, many are thrown down. And throughout the whole town there is a cry,—'*Par Dieu!* dame cat will never catch the rats [s].'"

Guillaume Guiart referring to the siege of Boves by Philip Augustus, speaks thus of the *cats* :—

"Devant Boves fit l'ost de France,
Qui contre les Flamans contance,

[s] *Hist. de la croisade contre les hérétiques Albigeois* (Hist. of the Crusade against the Albigenses), publ. by C. Fauriel, Collect. de docum. inédits sur l'hist. de France, 1re série, and the MS. of the Imp. Lib. (fonds La Vallière, No. 91). This MS. is by a contemporary, an eye-witness of the facts he relates; the exactness of the details gives this poem a great interest; we would particularly direct the attention of the reader to the description of the *gate* and of its progressing *by little leaps*, which affords a graphic image of the advance of these heavy pieces of framework carried forward on rollers, with sudden jerks. Such details as these, to be described so picturesquely, must have been seen.

Li mineur pas ne sommeillent,
Un chat bon et fort appareillent,
Tant eurent dessous, et tant cavent,
Qu'une grant part du mur destravent . . . [t]"

And in the year 1205 :—

" Un chat font sur le pont atraire,
Dont pieça mention feismes,
Qui fit de la roche meisme,
Li mineur desous se lancent,
Le fort mur à miner commencent,
Et font le chat si aombrer,
Que riens ne les peut encombrer [u]."

In order to protect the labourers who were making a causeway to cross a branch of the Nile, Saint Louis "caused to be made two towers (*baffraiz*), which are called *chas chateilz*. For there were two castles (*chateilz*) before the *cats* or galleries (*chas*), and two houses behind to receive the strokes which the Saracens dealt by their engines, whereof they had sixteen with which they did wonders [x]." The assailants supported their towers

[t] Before Boves was the army of France, which acts against the Flemings. The miners do not sleep, but prepare a *chat* good and strong; and so many get under it and so hard they work, that they destroy a great portion of the wall.

[u] A *cat* is drawn upon the bridge, which we have already mentioned as being a portion of the rock itself; the miners rush under it, commence to undermine the strong wall, and have the cat so well covered, that nothing can reach those within.

[x] The Sire de Joinville, *Hist. du roy Saint Louys*, edit. 1668. Du Cange, p. 37. In his observations, p. 66, Du Cange thus explains this passage :—" The king, Saint Louis, had therefore constructed two *beffrois*, or wooden towers, to guard those who were working at the causeway; and those towers were called *chats-chateils*, that is to say *cati castellati*, because above these cats (*chats*) there were castles. For these were not galleries simply, as the *cats* were, but galleries defended by towers and *beffrois*. Saint Louis, in one of his epistles, speaking of this causeway, says :—*Saraceni autem è contra totis resistentes conatibus machinis nostris quas erexeramus, ibidem machinas opposuerunt quamplures, quibus castella nostra lignea, quæ super passum collocari feceramus eundem, conquassata lapidibus et confracta combusserunt totaliter igne græco* . . . . And I believe that the lower story of these towers (*chateils*) were used as *cats* and galleries, wherefore the cats of this description were called *chas châtels*, that is to say, as I have already observed, cats fortified by castles. The author who has described the siege which was laid to Zara by the Venetians in the year 1346, lib. ii. c. 6, *apud Joan. Lucium de regno Dalmat.*, gives also an account of this kind of cat :—*Aliud erat hoc ingenium, unus cattus ligneus satis debilis erat confectionis, quem machinæ jadræ sæpius jactando penetrabant, in quo erat constructa quædam eminens turris duorum*

and cats by battering machines, catapults (*trébuchets*, *tribuquiaux*), mangonels (*mangoniaux*), *calabres* and pierriers, and by crossbow-men protected by boulevards or palisades filled in with earth and wattles, or by trenches, fascines, and mantelets. Those several engines (*trébuchets*, *calabres*, *mangonels*, and *pierriers*) were worked by counterpoise, and possesed great accuracy in their aim[y]; they could do no more, however, than destroy the crest-works and hinder the besieged from keeping upon their walls, or dismount their machines.

---

propugnaculorum. Ipsam duæ maximæ carrucæ supportabant. And because these machines were not simple *chats*, they were called *chats-faux* (*false cats*), being made in the form of turrets or towers, and nevertheless used as cats. And it is thus we should understand the following passage of Froissart:—'*On the day after, there came to the duke of Normandy two master-engineers, and said—Sir, if you will let us have timber and workmen we will make four great chaffaux* (some copies say *chats*) *which would be brought close to the walls and should be high enough to overtop them.*' Whence comes the word *eschaffaux* (scaffolds) amongst us, to signify a raised wooden platform."—See the *Recueil de Bourgogne*, by M. Pérard, p, 395.

y See *Etudes sur le passé et l'avenir de l'artillerie* (Studies of the Past and Future of Artillery), by Prince Louis-Napoleon Bonaparte, Presid. of the Republ., vol. ii. This work, characterized by much learned research, is certainly the most complete of all those on the subject of ancient artillery; the description given of the *trébuchet*, by the illustrious author, is as follows:—"It consisted of a beam called *verge* or *flèche*, turning round a horizontal axis supported upon uprights. At one extremity of the beam was fixed a counterpoise, and at the other a sling which contained the projectile. To make ready the machine, that is to say, to lower the *verge*, a winch was employed. The sling was the part of the machine of most importance, and according to experiments and calculations inserted by Colonel Dufour in his interesting Mémoire, *On the Artillery of the Ancients*, (Geneva, 1840) this sling increased the range to more than double, that is to say, that if the *flèche* had terminated merely *en cuilleron*, or in the form of a bowl, as was the case with certain hurling engines used by the nations of antiquity, the projectile, everything else being equal, would have been thrown to only half the distance it was with the sling.

"The experiments which we have made, on a small scale, have yielded the same result."

A machine of this kind was executed at full size, by the orders of the President of the Republic, in 1850, and tried at Vincennes. The *flèche* was 33.79 feet long, the counterpoise being fixed at 9,900 lbs.; and, after some preliminary experiments, a 24 pounder was hurled to a distance of 191 yards, a bomb of 0.22 m., filled with earth, to the distance of 155 yards, and bombs of 0.27 and 0.32 m. respectively, to 131 yards. (See the Report addressed to the Minister of War by Captain Favé, vol. ii. p. 38.)

From the earliest times the mine had been used to destroy portions of the walls and effect a breach. The miners, in so far of course as the nature of the ground permitted, cut a trench in the rear of the ditch, passed underneath, reached the foundations, sapped them and underpinned them with shores of timber covered with pitch and grease; then set fire to the shores and the walls fall. The besiegers, in order to protect themselves from these subterraneous works, established generally on the reverse of the ditch either palisades, or a continuous wall, the latter of which was a true covered-way which commanded the approaches, obliging the assailants to commence their mine at a considerable distance from the ditches; or, as a last resource, they countermined with the view of reaching the gallery of the assailants, whom, in that case, they drove back, or suffocated by throwing burning fascines into the galleries, and destroyed their works. There exists a curious Report of the seneschal of Carcassonne, Guillaume des Ormes, addressed to Queen Blanche, regent of France during the absence of Saint Louis on the raising of the siege laid to that place by Trencavel in 1240[z]. At that period the city of Carcassonne was not defended in the way we now find it[a]; the fortifications consisted merely of the Visigoth walls, repaired in the twelfth century, with a first enceinte (or lists), which could not have been of any great importance (see fig. 9), and some out-works (barbicans). The detailed bulletin of the operations of attack and defence of this place,

---

[z] See *Biblioth. de l'école des Chartes*, vol. vii. p. 363, Report published by M. Drouët d'Arcq. This text is reproduced in the work by Prince Napoleon Bonaparte (Napoleon III.), already quoted.

[a] Saint Louis and Philip the Bold executed immense works of fortification at Carcassonne, to which we shall have occasion to return.

given by the seneschal, Guillaume des Ormes, is in Latin, of which the following is a translation :—

"To the excellent and illustrious lady, Blanche, by the Grace of God Queen of the French, Guillaume des Ormes, seneschal of Carcassonne, her humble, devoted, and faithful servant, greeting :—

"Madam, these presents are to make known to your Excellency that the town of Carcassonne was besieged, by the self-called viscount and his accomplices, on Monday, Sept. 17, 1240. Whereupon we, who were in the place, instantly took from them the bourg (or suburb) of Graveillant, which lies in front of the Toulouse gate, and there we found great store of timber fit for carpentry uses, which was of great service. The said bourg extends from the city barbican to the angle of the said (fortified) place. The same day the enemy were enabled to carry a mill which we held, by reason of the multitude of people which they had (with them) [b]; and afterwards Olivier de Termes, Bernard Hugon de Serre-Longue, Géraut d'Aniort, and those who were with them, encamped between the angle of the town and the water [c]; and on the same day, by the aid of the ditches which were there, and by cutting up the roads lying between them and us, they shut themselves in, in order that we might not be able to go to them.

"On the other side, between the bridge and the castle barbican, were established Pierre de Fenouillet, and Renaud du Puy, Guillaume Fort, Pierre de la Tour, and many others of Carcassonne. At the one place and the other, they had so many crossbow-men that nobody could go out from the town.

"Afterwards they pointed a mangonel against our barbican; and we, on our side, forthwith planted in the barbican a Turkish *petraria* [d], exceeding good, which threw projectiles towards the said mangonel and around it; in such sort that, when they

---

[b] Probably the *moulin du roi* lying between the barbican of the castle and the river Aude.

[c] On the west side, see fig. 11.

[d] "Postea dressarunt mangonellum quemdam ante nostram barbacanam, et nos contra illum statim dressavimus quamdam petrariam turquesiam valde bonam, infra...."

wanted to fire upon us, immediately they saw the pole of our *petraria* in motion, they took to flight and abandoned altogether their mangonel; and there they formed ditches and palisades. We also, every time we fired off the *petraria*, retired from around it, because we could not go to them, on account of the ditches, fences, and wells which were there.

"Afterwards, Madam, they began a mine against the barbican of the Narbonnaise gate[e]; and forthwith we, having heard the noise of the work underground, made a counter mine, and we made in the inside of the barbican a great and strong wall of stones laid dry, so that we could thereby protect the full half of the barbican; and then they set fire to the hole they were making, in such wise that the wood having been burnt, a portion of the front of the barbican fell down.

"They began to mine against another turret of the lists[f]; we countermined and succeeded in taking possession of the chamber which they had formed. They began, thereupon, to run a mine between us and a certain wall, and they destroyed two crenelles of the lists; but we set up there a good and strong paling between them and us.

"They undermined also the angle of the town wall, in the direction of the bishop's house[g], and by dint of mining they

---

[e] To the east, see fig. 11.

[f] To the south, see fig. 11. The name of lists (*lices*) was given to an external wall or palisade of wood, formed beyond the walls, which formed a kind of covered-way; the lists were almost always protected by a shallow moat, and sometimes there was a second ditch between them and the town walls. By an extension of the term, the name of lists was given to the space comprised between the palisades and the town walls, and even to the external *enceintes*, when, at a later period, they were built of masonry and flanked by towers. The palisades which surrounded a camp were also called *lists*:—"Liciæ, castrorum aut urbium repagula." *Epist. anonymi de capta urbe CP.*, ann. 1204, apud Marten., vol. i. Anecd., col. 786: "Exercitum nostrum grossis palis circumcinximus et liciis." Will. Guiart MS.:—

> "... Là tendent les tentes faitices,
> Puis environnent l'ost de lices."

Le Roman de Garin:—

> "Devant les lices commencent li hustins."

Guill. archiep. Tyr. continuata Hist. Gallico idiomate v. 5. Ampliss. Collect. Marten., col. 620: "Car quant li chrestiens vindrent devant Alexandre, le baillif les fist herbergier, et faire bones lices entor eux, etc." (Du Cange, *Gloss.*)

[g] At the south-west angle, see fig. 11.

arrived under a certain Saracen wall [h], at the wall of the lists; but immediately as we perceived this, we made a good and strong paling between us and them, higher up in the lists, and we countermined. Thereupon they fired their mine and flung us to a distance of some ten fathom from our crenelles. But we forthwith made a good and strong paling and thereon a good brattish [i] (or breast-work) (15) with good *archières* [k]: so that none amongst them dared to come near us in this quarter.

"They began also, Madam, a mine against the barbican of the gate of Rhodez [l], and they kept beneath, because they wished to arrive at our walls [m], and they made a marvellous great passage; but we, having perceived it, forthwith made a great and strong paling both on one side and the other thereof; we countermined likewise, and having fallen in with them, we carried the chamber of their mine [n].

"Know also, Madam, that since the beginning of the siege

---

[h] *Saracen wall:* probably some out-work of the ancient Visigoth fortification.

[i] "*Bretachiæ*, castella lignea, quibus castra et oppida muniebantur; gallicè *bretesques, breteques, breteches*."—(Du Cange, *Gloss.*)

> "La ville fit mult richement garnir,
> Les fosses fere, et les murs enforcir,
> Les bretesches drecier et esbaudir."—(Le Roman de *Garin.*)

> "—As breteches monterent, et au mur quernelé . . .
> —Les breteches garnir, et les pertus garder.
> —Entour ont bretesches levées,
> Bien planchiées et quernelés."—(Le Roman de *Vacces.*)

. . . . The bretèches (in old English *Brettis, Brattish*), were often understood as *hourds* or hoards. The bretachial spoken of by the seneschal, Guillaume des Ormes, in his Report addressed to Queen Blanche, were temporary works erected behind the palisade to enable those within to attack the assailants after they had effected a breach. We give an illustration (15) of the works alluded to by the seneschal of Carcassonne.

[k] *Archières:* long and narrow slits made in the masonry of towers and fortified walls, or in hoarding and palisades, to allow arrows and bolts to be shot against the assailants.

[l] On the north, see fig. 11.

[m] This passage, as well as those which precede it, describing the mines of the besiegers, clearly proves that at that time the city of Carcassonne was provided with a double enceinte: the besiegers in fact are shewn to have passed under the outer enceinte for the purpose of undermining the inner rampart.

[n] Thus, when the besieged became aware of the miners being at work, they erected palisades both above and below the supposed opening of the gallery, in order to enclose the assailants between barricades which they were obliged to force before they could make any further advance.

Fig. 15. Part of Carcassonne defended by wood-work when a breach was made.

they have never ceased to make assaults upon us; but we had such good crossbows, and men animated with so true a desire to defend themselves, that it was in their assaults they suffered their heaviest losses.

"At last, on a certain Sunday, they called together all their men-at-arms, crossbow-men and others, and all, together, assailed the barbican, at a point below the castle [o]. We descended to the barbican, and hurled so many stones and bolts that we forced them to abandon the said assault, wherein several of them were killed and wounded [p].

---

[o] The principal barbican, situate on the side of the Aude, to the west, see fig. 11.

[p] In effect, it was necessary to descend from the castle situate on the crest of the hill, to the barbican, which commanded the faubourg lying at the base of the escarpment. See the plan of the city of Carcassonne, after the siege of 1240; fig. 16.

"But the Sunday following, after the feast of St. Michael, they made a very great assault on us; and we, thanks to God and our people, who shewed great good will in defending themselves, repulsed them; several amongst them were killed and wounded; none of our men, thanks be to God, were either killed or received a mortal wound. But at last, on Monday, Oct. 11, towards evening, they heard news that your people, Madam, were coming to our aid, and they set fire to the houses of the bourg of Carcassonne. They have destroyed wholly the houses of the Brothers Minors, and the houses of a monastery of the blessed Virgin Mary, which were in the bourg, to obtain the wood wherewith they made their palisades. All those who were engaged in the said siege abandoned it secretly that same night, even those who were resident in the bourg.

"As for us, we were well prepared, to God be thanks, to await, Madam, your assistance, so much so that none of our people were in want of provisions, how poor soever they might be; nay more, Madam, we had in abundance corn and meat enough to enable us to wait during a long time, if so it was necessary, for your succour. Know, Madam, that these evil doers killed, on the second day after their arrival, thirty-three priests and other clerks, whom they found on entering the bourg; know moreover, Madam, that the Seigneur Pierre de Voisin, your constable of Carcassonne, Raymond de Capendu, and Gérard d'Ermenville, have borne themselves very well in this affair. Nevertheless the constable, by his vigilance, his valour, and his coolness, distinguished himself above all others. As for the other matters concerning these parts, we will be able, Madam, to speak the truth to you respecting them when we shall be in your presence. Know therefore, Madam, that they had begun to mine against us strongly in seven places. We have nearly everywhere countermined, and have not spared our pains. They began to mine from the inside of their houses, so that we knew nothing thereof until they arrived at our lists.

"Done at Carcassonne, Oct. 13, 1240.

"Know, Madam, that the enemy have burnt the castles and the open places which they passed in their flight."

As for the battering-ram of the ancients, it was certainly employed in battering the base of the walls in sieges, from the twelfth century down. We borrow another passage from the Provençal poem of the Crusade against the Albigenses, a passage which can leave no doubt upon this point. Simon de Montfort wishes to succour the castle of Beaucaire which holds out for him, and is besieged by the inhabitants; he besieges the town, but has not constructed machines sufficient for his purpose; the assaults are without result; during this time, the people of Provençe press harder and harder on the castle (*capitole*).

". . . . But those of the town have raised against (the men of the castle) engines wherewith they batter the *capitole* and the watch-tower in such sort that the beams, the stone, and the lead are shattered by them; and on Easter-day is prepared the *bosson*, which *bosson* is long, straight, sharp, shod with iron; which so strikes, cuts, and smashes, that the wall is damaged, and several stones fly from it, here and there; but the besieged, when they perceive this, are not discouraged. They make a noose of rope, which is attached to a machine of wood, and by means thereof the head of the *bosson* is caught and held fast. Whereat the men of Beaucaire are greatly troubled, until there comes the engineer who set the *bosson* in motion. And many of the besiegers have planted themselves on the crags, to try and split the wall by striking it with sharpened pick-axes. And the men of the capitol, having perceived them, sew up, mingled together in a cloth, fire, sulphur, and flax, which they let down at the end of a chain alongside the wall, and when the fire has taken, and the sulphur melts, the flame and smell choke the pioneers to such a degree that not one of them can or does remain. But they go to their *petrariæ* (catapults) and make them play so well that they break and destroy the barriers and beams[q]."

This curious passage shews what were the means em-

---

[q] Passage of Provençal Poetry, p. 350.

ployed, in those times, for battering the walls at close quarters, when the object sought was to effect a breach, and the situation of the place did not allow of piercing galleries for the mines, placing shores under the foundations, or of setting fire to the works attacked. As regards the means of defence, there is mention at every page, in this history of the crusade against the Albigenses, of barriers, lists of wood, and palisades. When Simon de Montfort is obliged to return to besiege Toulouse, notwithstanding his having previously razed almost all the walls to their foundations, he finds the city defended by ditches and works constructed of timber. The castle called the Narbonnais, alone, is still in his power. The brother of the Count, Guy de Montfort, is the first to arrive with his terrible fanatics. The knights have dismounted, they break in the barriers and the gates, they force their way into the streets; but there they are received by the inhabitants and the men of the Count de Toulouse, and are forced to beat a retreat, when Simon arrives upon the scene, furious:—

"How comes it," he cries to his brother, "that ye have not, ere this, destroyed the town and burnt its houses?"

"We have attacked the town," replies Count Guy, "beaten in the defences, and found ourselves pell-mell with the inhabitants in the streets; there have we met knights and burghers, and workmen, armed with clubs, and bills, and sharp axes, who, with great shouts and hisses and deadly blows, have sent you, by us, your rents and your taxes, and Don Guy, your marshal, can tell you how many silver marcs they flung us down from their roof-tops! By the fealty I owe you, not a man among us is so brave, but, when they hunted us out through their gates, he would have liked better a fever or a pitched battle . . . ."

The Count de Montfort, however, is obliged to undertake a regular siege, after renewed and fruitless attacks.

"He posts his divisions (*batailles*) in the gardens, he furnishes the walls of the castle and the orchards with crossbows on wheels [r] and sharp arrows. On their side, the townsmen under their liege lord strengthen the barriers, occupy the grounds round about, and unfurl in divers places their banners, with two red crosses and the ensign of the Count (Raymond); whilst upon the scaffolds [s] and in the galleries [t] the most valiant and steady are posted, armed with poles shod with iron, and with stones to hurl down upon the enemy. Below, on the ground, have remained others, bearing lances and *dartz porcarissals*, to defend the lists, to the end that none of the assailants should be enabled to near the palisade. At the embrasures (*fenestrals*) and loop-holes archers defend the ambons and the parapets, with long bows of divers sorts, and hand crossbows. Tubs [u] are placed about full of bolts and arrows. Everywhere round about the crowd of people are armed with axes, clubs, poles shod with iron; whilst noble dames and the women of the city carry to them crocks and great stones, easy to hold and throw. The town is bravely fortified at its gates; bravely also and in good array do the barons of France, well stocked with fire and ladders, and heavy stones, draw near to the place from divers directions to seize on the barbicans [x] . . . ."

But the siege becomes protracted, and the winter comes; the Count de Montfort postpones his preparations for the attack until spring.

" . . . . Within and without are to be seen none but workmen, who fill the town, the gates and boulevards, the walls, the *brattishes* and the double palisades (*cadafalcs dobliers*), the moats, the lists, the bridges, the stair-flights. In Toulouse are none to be seen but carpenters, who make trebuchets and other engines, active and powerful, which in the castle of Narbonne,

---

r *Balestas tornissas* (verses 6,313 and foll.): probably as in the text.

s *Cadafals*, probably *bretachiæ*, see fig. 15.

t *Corseras*, hoards, or parapets probably; *coursières*.

u *Semals*. The wooden tubs in which the grapes are carried at harvest-time are still sometimes called *semals*, but more frequently *comportes*. They are oval-shaped, with wooden handles, through which two poles are passed for the purpose of carriage.

x *Bocals*, entrances to the lists.

against which they are pointed, leave neither tower, nor room, nor parapet, nor a whole wall, standing . . . ."

Simon de Montfort returns, he invests the town more closely, he seizes on the two towers which command the banks of the Garonne, he fortifies the hospital which lies outside the ramparts, and converts it into a fortress with moats, palisades, and barbicans complete. He strengthens his lines with sunk ditches, and walls pierced with several heights of embrasures. But after many an assault, and many a feat of arms devoid of good result to the besiegers, the Count de Montfort is killed by a projectile, launched from a *pierrier* worked by some women near to Saint-Sernin, and the siege is raised.

On his return from his first crusade, Saint Louis wished to make Carcassonne one of the strongest places in his *domaine*. The inhabitants of the faubourgs, who had opened their gates to the army of Trencavel[y], were driven out of their ruined dwellings, burnt by him whose cause they had espoused, and their ramparts were razed to the ground. It was not until seven years after this siege that Saint Louis, moved by the entreaties of Bishop Radulfus, granted by letters patent to the exiled burghers permission to rebuild a town on the opposite bank of the Aude, not wishing to have near the city subjects of such doubtful fidelity. The royal saint began by rebuilding the external enceinte, which was not of sufficient strength, and which had been much injured by the troops of Trencavel. He erected the enormous tower, called *la Barbacane*, as likewise the ramps which commanded the banks of the Aude and the bridge, and which allowed the garrison of the castle to make sorties

[y] The faubourgs which surrounded the city of Carcassonne were enclosed by walls and palisades at the date of the siege described by the Seneschal Guillaume des Ormes.

without being interrupted by the besiegers, were these even masters of the first lines. There is every reason for believing that the external towers and walls were erected somewhat hurriedly after the failure of Trencavel's expedition, in order first of all to place the city beyond the reach of a surprise, whilst the internal enceinte was being repaired and enlarged. The towers of this external line of wall were open towards the town, in order to render the possession of them useless to the besiegers, and the parapets of the curtains are on the same level as the ground of the lists, so that, if taken, they could not be used as a rampart against the besieged, who, being in force, would be able to throw themselves on the assailants and drive them back into the moats.

Philip the Bold, during the war with the King of Arragon, prosecuted these works with great activity until his death (A.D. 1285). Carcassonne was at that time a frontier place of great importance, and the King of France held his parliament there. He erected the curtains, towers, and gates, on the eastern side[z], advanced the internal line of fortification on the south side, and had the walls and towers repaired of the old Visigoth enceinte. We subjoin (fig. 16), the plan of the place as thus modified. At A is the great barbican on the side of the Aude, of which mention has been made, with its ramps fortified as far as the castle, F. These ramps, or slopes, are arranged so as to be commanded by the external defences of the castle; it was only after having passed through several gates, and followed various windings, that the assailant (admitting that he had obtained possession of the barbican) could arrive at the gate, L;

---

[z] Amongst others, the tower known as that of the *Trésau*, and the gate called the *Narbonnaise*.

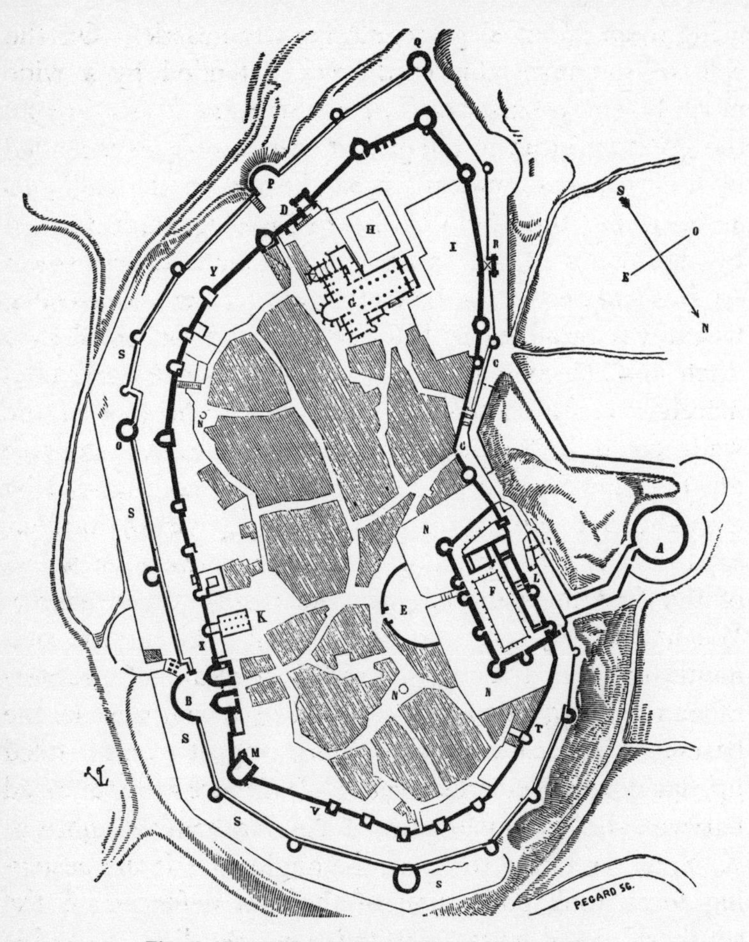

Fig. 16. Plan of Carcassonne as fortified by S. Louis.

A. The great Barbican. B. Gate of Narbonne. C. Gate of the Aude.
D. The great Postern. E. Barbican of the Castle. F. The Castle.
G. The Church. H. The Cloister. I. A Courtyard.
K. The Hall. L. Entrance Passage. M. The Tower of the Treasury.
N. The Moat of the Castle. O. A lofty Tower. P. Barbican of the Postern.
Q. Tower of the Angle. R. Square Tower. S. Ditch or Moat of the City.
T. V. X. Y. The Lists between the inner and outer walls of enceinte, or enclosure.

and here he would be obliged, within a narrow space completely commanded by towers and walls of great height, and having in his rear an escarpment which deprived him of all power of bringing up engines or of using them, to lay siege regularly to the castle. On the side of the town this castle was defended by a wide moat, N, and a barbican, E, built by Saint Louis. From the great barbican to the gate of the Aude you ascended by a narrow path, embattled on the side of the valley so as to defend the whole of the re-entering angle formed by the slopes of the castle and the walls of the town. At B is situate the Narbonnaise gate on the eastern side, which was provided with a barbican and protected by a ditch and a second barbican, the latter being palisaded merely. At S, on the side from which the foot of the walls could be reached almost on the level, is a wide moat. This moat and its approaches are commanded by a strong and lofty tower, O, itself an isolated fortress, capable, alone, of sustaining a siege, even were the whole of the first lines to fall into the hands of the enemy. We have every reason to believe that this tower communicated with the internal walls by means of a subterranean passage which was reached by a well sunk in the basement of the keep, but which, being at present filled up, has not yet been discovered. The lists are comprised between the two enclosures of the Narbonnaise gate at X, Y, as far as the tower at the angle, Q. If the besieging force took possession of the first defences on the south side, and wanted, by following the lists, to arrive at the gate of the Aude at C, he found himself stopped by a quadrangular tower, R, erected over and upon the two walls of the enceintes, and furnished with barriers and battlements. If he succeeded in passing between the Narbonnaise gate and the barbican at B,—a difficult task, —he had to cross over, in order to enter, at V, into the lists of the north-east, a narrow space, commanded by an enormous tower, M, called the *Tour du Trésau.* From V to T, he was taken in flank by the high towers of the

Visigoths, repaired by Saint Louis and Philip the Bold, and he found a further defence at the angle of the castle. At D is a great postern protected by a barbican, P; other posterns are distributed along the enceinte, allowing the guard to make the round of the lists, and even to reach the open country, without having to throw open the principal gates. This was an important point. It will be observed that the postern opening from the tower, D, and giving access to the lists, is placed laterally, and masked by the projection of the counterfort in the angle, the sill of this postern being more than two yards above the external ground level: thus it became necessary to plant a ladder against it for ingress or egress. From the numberless precautions then taken to defend the gates, it is natural to suppose that the assailants were in the habit of looking on them as weak points. The use of artillery has modified this opinion, by changing the means of attack; but at that time it may be conceived, that whatever may have been the obstacles accumulated round an entrance, the besiegers still preferred making an effort to overcome them, to establishing himself at the foot of a strong tower in order to undermine it by manual labour, or to batter it by means of engines of imperfect construction. Therefore when, during the twelfth, thirteenth, and fourteenth centuries, they wished to convey a high idea of the strength of a place, they said it had only one or two gates. But, as regarded the besieged, and especially when they had to guard a double line of fortifications, it was necessary nevertheless to have an easy communication between these two lines, in order to be able rapidly to send aid to any given point of attack. This is the reason why we find, in looking over the internal enceinte of the Carcassonne, a large number of posterns more or less concealed, and the purpose of which was to enable the garrison to spread themselves over the lists from several different points at the same given moment, or to fall rapidly back within the second walls, if the first had been carried. Beside the two great public gates, that of the Aude and the Narbonnaise, we reckon six open posterns, the internal levels of which are some yards above the outer ground-level, and access to which, consequently, could only be had by means of ladders. There is one, more particularly, pierced in the great curtain of the Bishop's Palace, which is only a little more than six feet high by somewhat less than a yard in width, and the sill of which is placed at a level of some thirty-nine feet above the ground of the lists. In the external enceinte, we discover another made in the curtain-wall between the gate of the Aude and the castle: this latter opens above an escarpment of rocks twenty-two or twenty-three feet high. Through these openings, in case of blockade and by means of rope ladders, the besieged could receive during the night emissaries from without without fear of treachery, or send forth into the country their messengers or spies. These two posterns, it will be seen, are placed on that side where the fortifications are inaccessible to the enemy by reason of the escarpment which overlooks the river Aude. The latter postern, which is pierced in the wall of the external enceinte, opens into the enclosed space protected by the great barbican and by the battlemented wall which followed the ramps of the river gate; it could thus serve, if need were, for pouring into this enclosure a company of determined soldiers, to cause a diversion, in case the enemy should be pressing too hard on the defences of this gate or on the barbican, by setting fire to the engines, turrets, or *chats*, of the besiegers. It is certain that a great importance was attached to the barbicans; for by these the besieged were enabled to make their sorties. On this ground the barbican of Carcassonne is of great interest (fig. 16). Built at the base of the hill, on the summit of which the castle is erected, it places this latter in communication with the banks of the Aude [a]; it forces the assailant to keep at a distance from the ramparts of the castle; of dimension vast enough to contain from fifteen to eighteen hundred foot-soldiers, without reckoning those which manned the walls, it permitted the concentration of a considerable body of troops who might, by a vigorous sortie, drive back the besiegers into the river. The barbican of the castle of the city of Carcassonne completely masks the gate, B, the slopes of which descend into the open country. Those slopes, E, are battlemented right and left. Their continuity is broken by pierced screen walls, and the whole work, which ascends by a steep ascent towards the castle, is enfiladed in its entire length by a tower and two upper curtain walls. If the besiegers succeeded in arriving at the top of the first slope, they had to make a detour in the direction E′, and were then taken in flank; at F they found a fortified parapet, and further, a gate strongly defended and embattled; if they got through this first gate, they had to pass along a parapet with pierced embrasures, force a barrier, then take a sudden turn and attack a second gate, G, exposed also to a flank fire. Having taken this, they found themselves before a considerable and well-defended work: this consisted of a long passage surmounted by two

---

[a] The plan which we here give is drawn to a scale of 1 *centimètre* to 15 *mètres*. The barbican of Carcassonne was destroyed in 1821, to allow of the erection of a mill; its foundations only exist, but its ramparts are for the most part preserved, particularly in the portion adjoining the castle, which is the most interesting.

stories, under which they had to pass. The first commanded the last gate, by means of a defence of timber-work, and was pierced with loop-holes the whole length of the passage; the second communicated with the pierced parapet opening either on the exterior on the side of the slopes, or on the portion over this same passage. The floor of the first story communicated with the parapet of the outer walls of the lists only by means of a small gate. If the assailants succeeded in obtaining possession of it by escalade, they were taken in a trap; for, once the small gate closed upon them, they found themselves exposed to the projectiles flung from the battlements of the second story, and the extremity of the floor being suddenly cut off at H on the side opposite the entrance, they found it impossible to advance beyond. If they cleared the passage on the ground-level, they were stopped by the third gate, H, in a wall surmounted by the battlements of the third story, communicating with the upper parapet-walk of the castle. If, by an almost impossible chance, they were able to seize on the second story, they found there no other mode of egress than through a small gate opening into a second room situate along the walls of the castle, and whose only means of communication with the latter was by winding passages which could be easily barricaded in a moment, and which, moreover, were defended by strong embrasures. If, in spite of all these accumulated obstacles, the besiegers forced the third gate, they then would have to attack the postern, I, of the castle, guarded by a formidable system of defences: loop-holes, two rows of battlements, one over the other, a drawbridge, a portcullis and some embrasures. Were this gate carried, they were still at a depth of more than seven yards below the internal courtyard, L, of the castle, which they could reach only along narrow sloping passages, and by passing through several doors at K.

Supposing the attack was made from the side of the gate of the Aude, they were stopped by a guard-house, T, by a gate with a work of timber, and by a double row of battlements pierced in the floor of an upper story, communicating with the great south hall of the castle by means of a passage constructed of wood, which could be destroyed at a moment's notice; so that, by taking possession of this upper story, you had effected nothing. If, after having forced the gate on the ground-floor, you pushed on in advance along the parapet-way of the great square tower, S, you came shortly upon a gate strongly defended with battlements, and built parallel to the passage, G H. Beyond this gate with its defences was a second gate, narrow and low, in the massive inner wall, Z, which had to be forced; then, finally, you arrived at the postern, I, of the castle. If, on the contrary (but this was an impossible undertaking), the assailant presented himself on the opposite side, by the north lists, he was stopped by a defence at V. On this side, however, the attack could not be attempted, for this is the point of the city which is most strongly defended by nature; and in order to force the first enceinte between the *Tour de Trésau* (see fig. 16) and the angle of the castle, it was necessary first of all to climb a steep incline, and to scale the crags. Besides, in attacking the north gate, V, the besiegers presented themselves in flank before the defenders who manned the high walls and towers of the second enceinte. The massive internal wall, Z, which, starting from the curtain of the castle, advances at right angles as far as the descent to the barbican, was crowned by transverse battlements which commanded the gate, H, and was terminated at its extremity by a bartizan, or watch-turret, which allowed what was passing in the *rampe*, or sloping walk, descending to the barbican, to be seen, with a view to internal measures of defence in case of surprise; or to reconnoitre the troops when returning from the barbican to the castle.

The castle could thus hold out for a long time, although the town and its environs should be in the hands of the enemy; its garrison, defending with ease the barbican and its ramparts, remained masters of the Aude (the bed of which was at that time closer to the city than it is at the present day), and could thus receive their provisions by the way of the river, and prevent any blockade from being effected on that side; since it was quite impossible for a body of troops to take up a position between this barbican and the Aude without danger, they having no cover, and the flat and marshy ground being commanded on all sides. The barbican had the further advantage of placing the king's mill in communication with the garrison of the castle, and this mill was itself fortified. A plan of the city of Carcassonne, surveyed in 1774, mentions in its title a great subterraneous passage existing under the boulevard of the barbican, but which had been long closed up and partially filled. This subterraneous passage may have been intended to establish a secret communication between this mill and the fortress.

On the side of the town, the castle of Carcassonne was likewise defended by a great barbican, C, in advance of the moat. A gate, A′, strongly defended, gave entrance into this barbican; the bridge, C, communicated with the principal gate, O. Vast porticoes, or sheds, N, were provided for lodging a temporary garrison in case of siege.

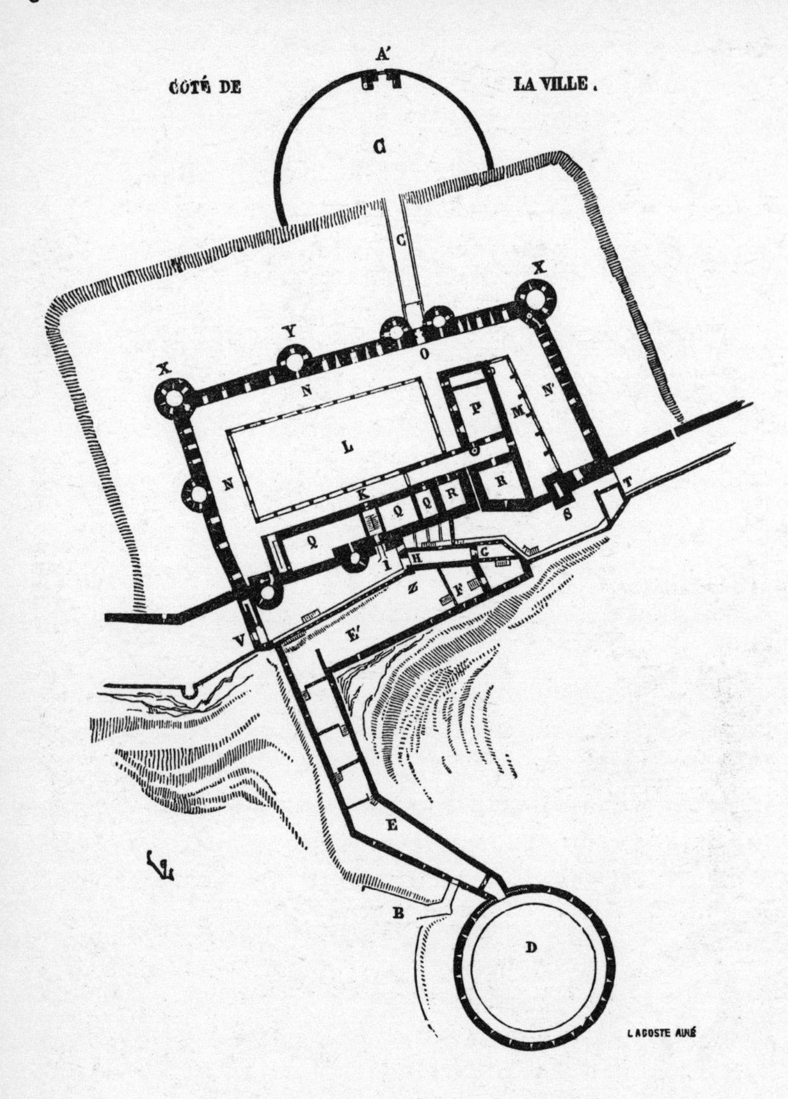

Fig. 17. **Plan of the Castle of Carcassonne.**

A. The outer Gate from the Barbican towards the City.
B. Outer Gate of the Castle to the Country.
C. The Bridge across the Moat.
D. The outer Barbican.
E E'. Passage from the outer Barbican to the Castle.
F. Parapet to protect the Gate.
G. Second Gate.
H. Third Gate
I. Postern of the Castle.
K. Passage.
L. Principal Courtyard.
M. Small Courtyard
N N N. The Porticoes.
O. Principal Gate.
P & Q. Barracks.
R R. The Keep Towers.
S. Great Square Watch-tower
T. Guard-house.
V. The North Gate.
X Y. Towers.
Z. Tower Wall.

Fig. 18.

For the ordinary garrison, quarters were provided in buildings of three stories, Q, P, situate beside the Aude. Over the portico, N′, south side, was a vast *salle d'armes*, with pierced loop-holes on the side of the moat, and with windows opening upon the courtyards. R R were the keeps, the greatest of these being separated from the neighbouring buildings by an open space, and only communicating with these latter by means of wooden bridges, easily removed. Thus, the castle being taken, those of the garrison who were left could still find a refuge within this enormous tower, which was completely closed, and hold out for a time. At S is an immense watch-tower which overlooks the entire town and its environs ; a wooden staircase was all that it contained. The towers, X, Y, the gate, O, and the intermediate curtain-walls are of the twelfth century, as likewise the watch-tower, and the basement of the buildings on the side of the barbican. These structures were completed and restored under Saint Louis. The great barbican of the Aude had two heights of loop-holes and an upper embattled parapet-walk, which could be furnished with hoarding.

We give (fig. 18) a bird's-eye view of this castle and its barbican, which, taken with the plan (fig. 17), will complete the description we have just given of it ; it will be easy to mark out the position of each portion of the defences. We have supposed the fortifications as in a complete state of defence, and provided with all their war accessories, of wooden defences, brattishes, hoarding, and advanced palisades.

But it is requisite, before proceeding further, to explain fully what these *hourds*, hoarding, or hoards, were, and the motives which led to their adoption, from the twelfth century downwards.

The danger of defences of wood on the ground-level had been discovered; the assailant could easily destroy them by fire; and as early as the time of Saint Louis, the wooden lists and barbicans, so frequently employed in the preceding century, were replaced by external walls (enceintes) and by barbicans built of masonry. They did not, however, entirely abandon the use of timber defences, but took care they should be placed at such a height as would render their destruction, by means of incendiary projectiles, difficult at least, if not impossible. At that time as now (and the fortifications of the city of Carcassonne are there to furnish an example), when good defences were required, care was taken to secure everywhere above the ground-plane which lay at the foot of the walls and towers a *minimum* of height, in order to place them all equally beyond the reach of escalade, along their whole line. This *minimum* height is not the same for the two lines of defence, the inner and outer enceintes; the curtain-walls of the first defence are maintained at a height of about thirty-three feet from the bottom of the moat, or from the crest of the escarpment to the floor of the hoards, whilst the curtain-walls of the second enceinte are, from the ground-level of the lists to the floor of the hoards, forty-seven feet at least. The ground which forms a plateau for the two enceintes not being horizontal, but presenting, on the contrary, very considerable differences of level, the ramparts follow the natural slopes of the ground, and the hoards conform to the inclines of the parapet-way. There were thus, as we see, at that time certain data, rules, and formulæ for military architecture, in the same way that there were similar rules for religious and civil architecture. The remaining portion of this article will furnish superabundant proofs, we consider, of this fact.

According to the system of battlements and loopholes, or eyelets, pierced in stone parapets, it was not possible to hinder a force of assailants, when bold and numerous, and protected by *chats* covered with skins or cushions, from undermining the foot of the towers or curtain-walls, inasmnch as it was impossible from the loopholes, notwithstanding the inclination of their sectional line, to see the foot of the fortifications; nor was it possible to take aim through the battlements, without at least projecting one half of the body beyond the line of wall, at any object at the base. It became necessary, therefore, to construct projecting galleries, well provided with defences, and which would allow a large number of the besieged to overhang the base of the wall, so as to be able to hurl down on an attacking party a perfect hail of stones and projectiles of every kind. Let fig. 19 be a curtain-wall crowned by a parapet with battlements and loopholes, the man placed at A cannot see the pioneer, B, except on the condition of advancing his head beyond the battlements; but in that case he completely uncovers himself, and whenever pioneers were sent forward to the foot of a wall, care was taken to protect them whilst at work by discharging showers of arrows and cross-bolts wherever the besieged were visible. In time of siege, from the date of the twelfth century[b], the parapets were provided with hoards, C, in order to command completely the base of the walls by means of a continuous machicolation, D. Not only did the hoards perfectly accomplish this object, but they left the defenders entirely free in their movements, as the bringing

[b] The castle of the city of Carcassonne is of the commencement of the twelfth century, and all its towers and curtain-walls were well supplied with hoards, which must have been of great projection, from the precautions taken to prevent the sagging (*bascule*) of the timbers of the floor.

up the supplies of projectiles and the circulation was carried on behind the parapet at E. Further, when

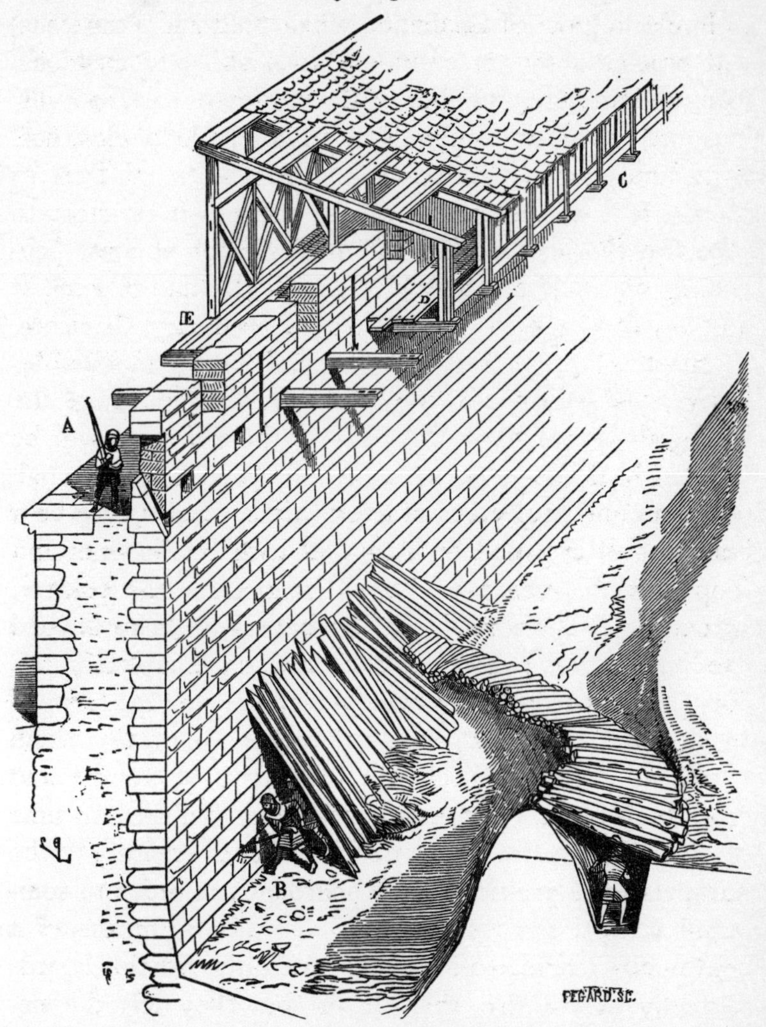

Fig. 19. A Curtain-wall with Battlement and Loopholes; and the Wood-work shewing one mode of attack and defence.

A. A Guard. B. A Pioneer. C. The Hoarding. D. The Machicoulis. E. The Platform for a Passage inside the Parapet.

these hoards were constructed, besides the continuous machicolation, with loopholes, the archères, or arrow-slits, formed in the masonry remained uncovered at their lower extremity, and allowed the archers and crossbow-men, who were posted within the parapet, to fire upon the assailants. With such a system the defence was as active as possible, and nothing but the lack of projectiles could afford any respite to the besiegers. We must not therefore feel surprise if, during some memorable sieges, after a prolonged defence, the besieged were reduced to the necessity of tearing the roofs from their houses, demolishing the walls of their gardens, and taking up the pavement of the streets, in order to keep the hoards supplied with projectiles, and thus force the assailants back from the foot of the fortifications. These hoards were readily and easily placed in position; in times of peace they were removed. We subjoin the representation (fig. 20) of the works of approach of a curtain-wall flanked by towers and with wet moat, in order to render intelligible the several means of defence and attack to which we have alluded. In the foreground is a cat, A; this is used to fill up the moat, and advances towards the foot of the wall upon the heaps of fascines and materials of every kind which the assailants are constantly engaged in flinging before them, through an opening in front of the cat; a wooden boarding which is fixed as the cat advances allows of its being moved along without any risk of its sticking fast in the mud. This engine is propelled either by rollers in the inside worked by levers, or by cords and fixed pulleys, B. In addition to the shed which is placed in front of the cat, palisades and moveable mantelets protect the labourers. The cat is covered with raw hides, in order to preserve it from the inflammable materials which may be launched by the besieged. The assailants, before sending the cat forward against the curtain-wall for the purpose of undermining

Fig. 20.

A. The Cat. B. The Pulley. C. The Catapult.
D. The Crossbow-men. E. The Wooden Tower and Drawbridge.

its base, have destroyed the hoards of this curtain-wall by means of projectiles, thrown by their slinging machines. Further on, at C, is a great catapult; it is directed against the hoards of the second curtain. This engine is ready strung; a man places the sling with its stone in position. A lofty palisade protects the engine. Close by are crossbow-men behind rolling mantelets, who take aim at any of the besiegers who leave their cover. Beyond these, at E, is a turret furnished with its moveable bridge, covered with hides: it advances upon a prepared floor, the boards of which are laid down according as the assailants, protected by palisades, fill up the moat; it is moved, like the cat, by ropes and fixed pulleys. Still further is a battery of two catapults, which are hurling barrels filled with incendiary material against the hoards of the curtain-walls. Within the town, upon a great square tower terminating in a platform at the summit, the besieged have fixed a catapult which is directed against the turret of the assailants. Behind the walls another catapult, covered by the curtains, hurls projectiles against the engines of the assailants. So long as the machines of the enemy have not arrived at the foot of the the walls, the part played by the besieged is almost passive; they content themselves with launching through the loop-holes of their hoards as many arrows and bolts as they can, If they are bold and numerous, they may attempt in the night to fire the turret, the palisades, and machines, by issuing from some postern at a distance from the point of attack; but if timid or demoralised, if they have no bold and devoted band amongst their ranks, at daybreak their moat will be filled, the floor of planks slightly inclined towards the walls will allow the turret to advance rapidly by its own weight, and the assailants will have but to maintain it in its place. Upon the fragments of the hoards crushed

Fig. 21.

by the stones hurled from the catapults, the moveable bridge of the turret will suddenly descend, and a numerous troop of knights and picked soldiers will precipitate themselves upon the parapet-way of the curtain (fig. 21). But this catastrophe is foreseen: if the garrison be faithful, abandoning the taken curtain, they will shut themselves up in the towers which are placed at intervals along it (fig. 22 [c]); there they can rally, enfilade the parapet-walk and cover it with projectiles, and, through the two gates, A and B, make a sudden sortie while the assailants are endeavouring to descend into the city; and before they have become too numerous, drive them back, seize upon the turret, and set it on fire. If the garrison, driven back, are incapable of so bold a stroke, they barricade themselves in the towers, and the assailants will have to besiege each of them in turn; for, where need is, every tower can be turned into a small independent fort, and many of them are provided with wells, ovens, and cellarage for storing

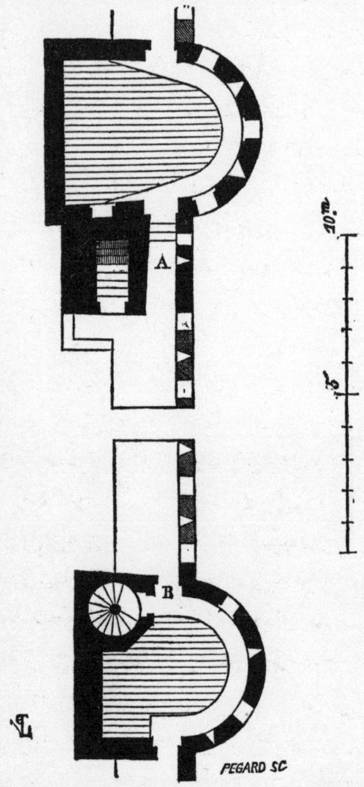

Fig. 22. Plan of one Bay of the Curtain-wall, and two Bastions or Towers, Carcassonne.

A & B. Doors from the Bastions to the Alure over the Wall.

[c] The example here given is taken from the interior enceinte of the city of Carcassonne, in that part built by Philip the Bold. The plan of the towers is taken at the level of the curtain; and is that of those known as the Daréja and Saint Laurence towers, south side.

and cooking provisions. The gateways by which the towers communicate with the parapet-walk are narrow, iron-plated, closed on the inside, and strengthened by wooden bars let into the thickness of the wall, in such a way that in a moment the door can be drawn to and rapidly barricaded by inserting the wooden bar.

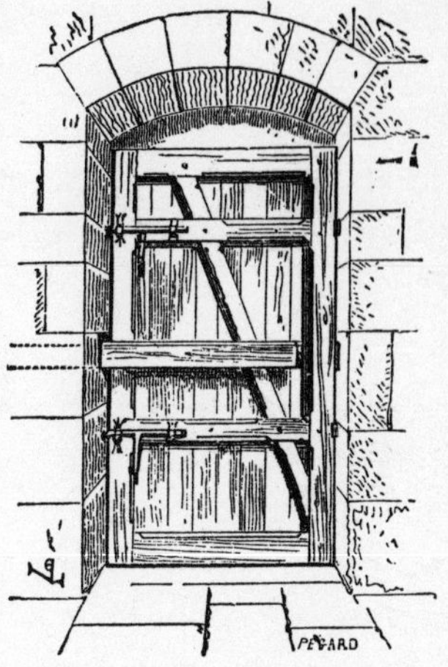
Fig. 23. Wooden Door of a Bastion.

We are struck, when we study the system of defence adopted from the twelfth to the sixteenth century, with the care taken to guard against surprise; all kinds of precautions are taken to arrest the progress of the enemy and to embarrass him at every step by complicated arrangements in the plan, and by turns and checks which it was impossible he could foresee. It is evident that a siege, before the invention of cannon, was never really serious, either for the besieged or the assailants, except when it became a hand-to-hand contest. A veteran garrison could still struggle on, and with some chance of success, until driven to their last defences. The enemy might enter the town by an escalade, or by a breach, without the garrisons being on that account forced to surrender; when this occurred, shut up in their towers, which, we repeat, were so many separate forts, they could make a long resistance, exhaust the strength of the enemy, and cause him heavy losses in every partial attack; for many a well-barricaded doorway had to be broken in, and many a stout conflict to be sustained, hand-to-hand, within spaces circumscribed and encumbered. Should the ground-floor of a tower happen to be taken, the upper stories had still powerful means of defence. We see that everything was calculated beforehand for disputing the ground foot by foot. The spiral staircases which gave access to the various stories of the towers were easily and promptly barricaded, in such a way as to render hopeless all efforts of the assailants to ascend from one story to another. Even if it happened that the burghers of a city wished to capitulate, the garrison might still hold out against them, and debar them from all access to the towers and curtains. It was a system of universal suspicion.

It is in all these details of foot-by-foot defence that the art of fortification, as pursued from the eleventh to the sixteenth century, appears in its best aspect. It is by carefully examining every trace that remains of the defensive obstacles of these times that we are enabled to understand those narratives of gigantic attacks, which we are too frequently disposed to tax with exaggeration. Whilst attentively considering these means of defence, so ably thought over and combined beforehand, we can easily figure to ourselves the immense labours of the besiegers, their moveable turrets, their contravallations, their boulevards, their bastilles, and all the various means of attack which were brought into play against a beleaguered enemy, who himself had calculated every chance of assault, who frequently acted on the offensive, and who was never disposed to yield a foot of ground, unless he could retire to another position stronger than the one he quitted.

At the present day, thanks to our artillery, a general who invests a fortified place which is not supported by an army in the field, can foretel the day and the hour when that place will fall. He can tell beforehand the moment when his breach will be practicable, or when his columns of attack will enter a given work. It is a game which takes more or less time in the playing, but which the besieging party are always certain to win, if there be no lack of ammunition, if they have an army at their disposal proportioned to the force of the garrison, and skilful engineers. "Place attaquée, place prise" (a place attacked is a place taken), says the French proverb[d]. But at that time nobody could tell when or how a fortified place would fall into the hands of the besiegers, how numerous soever they might be. With a determined garrison and plenty of provisions, a siege might be prolonged indefinitely: nor was it rare to see a *bicoque* resisting during many months a numerous and veteran army. Hence arose, frequently, that boldness and insolence of the weak towards the powerful; that habit of individual resistance which constitutes the ground-work of the character of feudalism; that energy which gave birth to such mighty things in the midst of so many abuses, which enabled the French and Anglo-Norman peoples to recover themselves after terrible reverses, and to found enduring nationalities; and by means of which they ever discovered unknown resources when their fortunes seemed at the lowest ebb.

---

[d] Like many others, this proverb is not altogether true, however, and many examples occur to gainsay its accuracy. It is certain that, even at the present day, a place defended by a commander of ingenuity and skill, one whose coup-d'œil is rapid and accurate, may hold out much longer than one which is defended by a man of routine whose intelligence cannot furnish fresh resources at every phase of the attack. It may be found, perhaps, that since siege warfare has become a science, and a kind of formula, as it were, we have been led to make too light of all those resources of detail which were still in use down to the sixteenth century. We cannot doubt but that the study of archæology, which has had so great an influence over the other branches of architecture, will exert its action upon military architecture likewise; for in our opinion (and it is one shared by competent authorities), although there may be nothing in the *form* of the fortification of the middle ages of which use might be made at the present day, owing to the powerful agency of artillery, the same is not true of its *spirit*, or of its principles.

Nothing is better adapted to bring into strong relief the profound differences which separate the characters of the men of those remote ages from the spirit of our own time, than to institute a comparison between a fortified city or castle of the thirteenth or fourteenth century and a modern fortress. In this latter there is nothing to strike the eye, everything wears an uniform appearance, and one bastion so closely resembles another that it is hard to recognise any one individually. An army advances against a city and takes it, yet the besiegers have scarcely seen the besieged; for weeks and weeks they have seen nothing before them save some heaps of earth and a little smoke. The breach is made and the place capitulates; everything falls the same day; a piece of wall has been thrown down, a little earth dislodged, and the city, the bastions, which have not seen even the smoke of the guns, magazines, arsenals, everything is given up. Humanity, considered in its material aspect, is a gainer; for the immediate disasters, the fury and the excesses which follow in the wake of a successful assault, are avoided: but the sentiment of responsibility and of individual resistance is lost, the energy of national character is enfeebled. Some hundreds of years ago things were differently managed. If a garrison were faithful, and good soldiers, it was necessary, so to speak, to force every tower to capitulate, to treat with every captain who was minded to defend, foot by foot, the post which had been confided to him. Everything, at least, was arranged with a view to this result. People accustomed themselves to rely only on their own powers and that of those about them, and they defended themselves against all comers. In this way (for we may deduce the greater from the less) it was not in those days enough to take the capital of a country for the country to be yours. Times of barbarism they may have been, but it was a barbarism full of energy and resources. The study of these great military monuments of the middle ages is, therefore, not curious only, it gives us an insight into habits of thought and action to which our national character might do well to return.

We behold at the beginning of the thirteenth century the inhabitants of Toulouse, with some great lords and their knights, in a badly enclosed city holding in check the army of the powerful Count de Montfort, and forcing him to raise the siege. But it was not the cities alone which thus acted; the great vassals, shut up in their castles, were at all times ready to resist, not only their rivals, but even their suzerain and his armies.

"The distinctive, the general character of feudalism," says M. Guizot, "is the dismemberment of the people and power into a multitude of little peoples and small sovereigns; the absence of any general nation, or any central government. . . . . Under what enemies did feudalism fall? Who struggled against it in France? Two forces: royalty on the one hand, the communes on the other. By means of royalty a central government came to be established in France; by the communes a general nation was formed, which in time grouped itself round the central government[e]."

The development of the feudal system is, therefore, limited to the period between the tenth and fourteenth centuries. It was then that feudalism erected its most important fortresses; that it completed, during the struggles of baron against baron, the military education of the nations of the West.

"With the fourteenth century," adds the illustrious historian, "wars change their character. Then begins the series of foreign wars; no longer wars of vassal against suzerain or of vassal

---

[e] Hist. of Civilization in France, by M. Guizot, 2nd part, 1st lesson.

against vassal, but of one people with another, of sovereign against sovereign. At the accession of Philip de Valois, the great wars of the French against the English break out; the claims of the kings of England, not upon this or that fief, but upon the country and the throne of France, are put forward; and these wars last until the time of Louis XI. It is therefore no longer feudal wars which are in question, but national wars; a certain proof that the feudal epoch stops at these limits, and that another social state has already begun."

But without feudalism, without the trials to which the nation had been subjected under its sway, and which were imposed by its very nature, could France have struggled during more than a century with her enemies from the other side of the channel, have waged battle at one and the same time against both foreign and domestic enemies, have preserved her national character, and been as strongly constituted the day after as the day before the contest? However barbarous and oppressive the feudal system may now appear, we consider it entitled to this tribute. To it we owe our best activity and strength; and the very men who, at the close of the last century, overthrew its last vestiges, would not have found in the country the energy which is its traditional characteristic, had the nation not been brought up in this hard school. It may be as well to bear this in mind.

The feudal castle is invested with its true defensive character only when it is isolated, at a distance from great, wealthy, and populous cities, and when it overawes some little town, village, or hamlet. It then takes every advantage of the configuration of the country, and surrounds itself with precipices, moats, and watercourses. When it forms a part of the city, it becomes its citadel, and is obliged to keep its defences subordinate to those of the city walls, to place itself at some point from which it can remain master of the parts beyond its walls and within them. In order to convey our meaning in a few words, we may say, that the true feudal castle, viewed with reference to the art of fortification, is that which, having itself fixed upon its site, sees by degrees the habitations of the people gradually come and group themselves around it. Far other is the castle whose construction, being of later date than that of the town, has found itself obliged to make its site and arrangements dependent upon the situation and the defensive arrangements of the city. At Paris, the louvre of Philip-Augustus was evidently constructed in accordance with the latter conditions. Until the reign of that prince, the kings of France inhabited ordinarily the palace which was situate in the city. But when the city of Paris had assumed a considerable development on both banks of the river, this central residence could no longer be a suitable one for the sovereign, whilst as a defence it had become quite useless. Philip-Augustus, in building the louvre, planted a citadel at the point of the city where he had most to fear from attack, and where his formidable rival, Richard, was most likely to present himself; he kept guard over both banks of the Seine above the city, and commanded the marshes and the fields which, from this point, at that time extended to the slopes of Chaillot and as far as Meudon. When he enclosed the town by walls, he took care to leave his new castle, his citadel, outside of their limits, in order thereby to preserve its liberty of defence. We see in the plan of Paris (fig. 24), as we have already observed, that besides the louvre, A, other fortified establishments are scattered around the walls; H is the Château du Bois, surrounded by gardens, a pleasure residence of the king's. At L is the hôtel of the dukes of Brittany. At C the palace of King Robert, and the monastery of Saint Martin-of-the-Fields, surrounded by a fortified enceinte. At B the temple, forming a separate citadel, with its walls and keep. At G

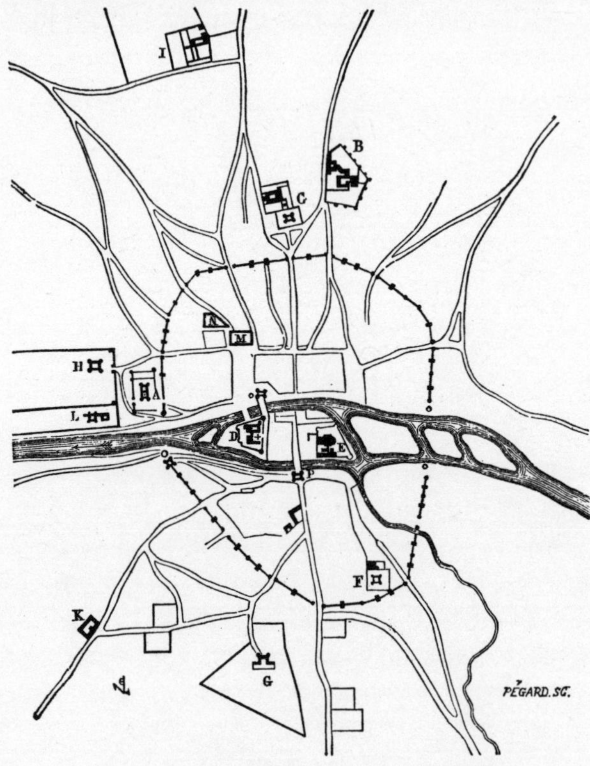

Fig. 24. Plan of Paris in the Thirteenth Century.

A. The Louvre.
B. The Temple.
C. Palace of King Robert.
D. The Law Courts.
E. Notre Dame.
F. Saint Geneviève.
G. Hotel de Vauvert.
H. Chateau du Bois.
I. House of S. Lazare.
K. The Infirmary.
L. Palace of the Duke of Brittany.
M & N. The Markets.
O. The Grand-Chatelet.
P. The Petit-Chatelet.

the Hôtel de Vauvert, built by King Robert, likewise surrounded by an enceinte[f].

---

[f] At I was the house of Saint-Lazare; at K the infirmary; at M and N the markets; at O the grand-châtelet, which guarded the entrance to the city from the north, and at P the petit-châtelet, which defended the Petit-Pont, on the south; at E Nótre Dame and the bishop's palace; at D the ancient courts of law; at F Sainte-Geneviève and the palace of Clovis, on the mountain.—(*Descript. de Paris*, par N. de Ferè, 1724.)

At a later period, during the imprisonment of King John, it was found necessary to extend the city walls. The city growing larger and larger, especially on the side of the right bank (fig. 25), the Louvre and the Temple became

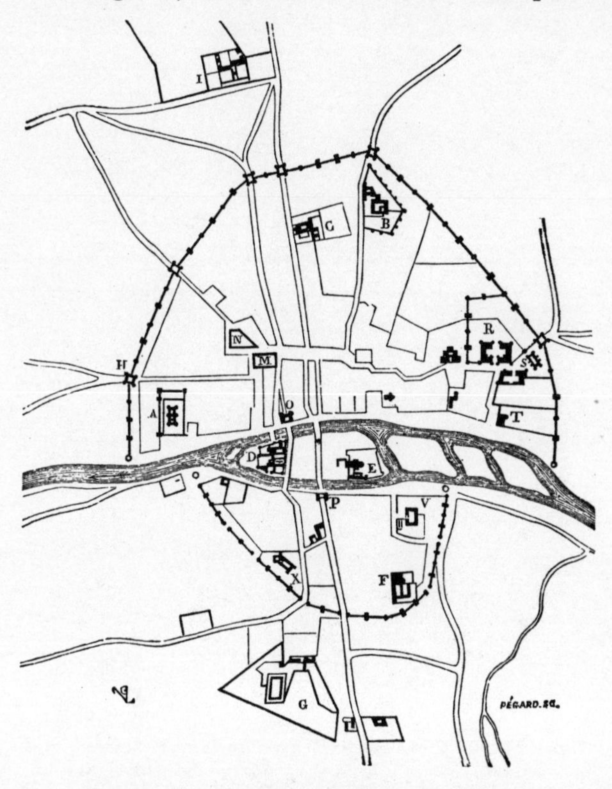

Fig. 25. Plan of Paris in the Fourteenth Century.

A to P. The same as in the Thirteenth Century, (see opposite). R. Palais des Tournelles. S. Bastille of S. Antoine.

enclosed within the new walls; but gates of good defence, and provided with barbicans, served the purpose of detached forts, and on the eastern side Charles V. had the bastille of Saint Antoine, S, erected, which commanded the faubourgs, and served as a support to the enceinte. The Palais des Tournelles, R, gave further strength to this portion of the city; and moreover, the Temple and the

Louvre, which preserved their fortified walls of enclosure, formed, in conjunction with the Bastille, so many internal citadels. We have already mentioned that the system of fortification adopted in the middle ages was not adapted to extended lines of defence; its force became impaired when the circumference of these became too great, unless they were accompanied by those advanced fortresses which divided the forces of the besiegers and impeded the advance of an enemy. We have seen in the case of Carcassonne (fig. 16) a town of small dimensions well defended by art and by the nature of the ground; but the castle there forms a portion of the city, is no more than its citadel, and has none of the characteristics of a feudal castle; while at Coucy, for example (fig. 26), although the castle is annexed to the town, it is completely independent of it, and preserves its character of a feudal castle. Here the town, built at C, is surrounded by an enceinte of considerable strength; between it and the castle, B, lies an esplanade or kind of *place d'armes*, A, communicating with the town only by the gate, E, with defences on both sides, but more especially against the town. The castle, built on the crest of the hill, looks down over very steep escarpments and is separated from the place d'armes by a large moat, D. If the town were taken, the place d'armes, and, behind it, the castle, served as secure places of refuge

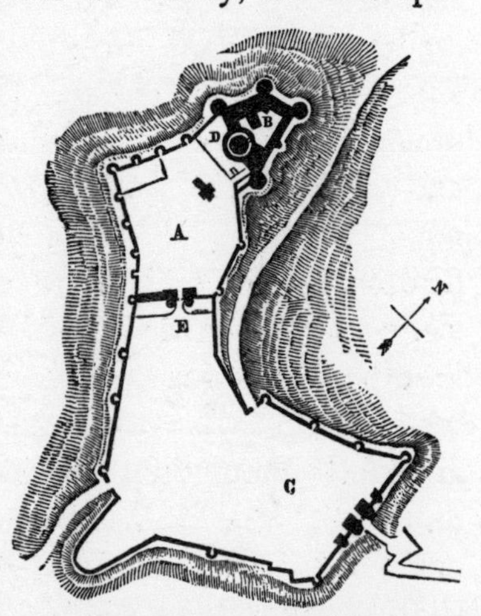

Fig. 26. Castle of Coucy.

A. Place d'Armes. B. The Castle.
C. The Town. D. The Moat.
E. The Gate from the Town to the Castle.

for the garrison. It was in the space, A, that the stables, household offices, and barracks of the garrison were placed, so long as they were not obliged to retire within the lines of the castle; sally-ports in the curtain of the place d'armes allowed of their making sorties, or of receiving assistance from without if the enemy held the city and was not in force sufficient to invest it and blockade the castle. Several towns offered dispositions of defence analogous to those here described:—Guise, Château-Thierry, Châtillon-sur-Seine, Falaise, Meulan, Dieppe, Saumur, Bourbon-l'Archambaut, Montfort-l'Amaury, Montargis, Boussac, Orange, Hyères, Loches, Chauvigny in Poitou, &c. In this latter city, three castles commanded the town at the close of the fourteenth century, all three built upon a neighbouring hill, and all three independent of each other. Those cities in which the defences were thus divided were considered, and justly, as of great strength; frequently hostile armies, after having taken possession of the town fortifications, were obliged to relinquish attempting to lay siege to the castle; and following up their conquests in other directions, left garrisons behind them intact, who, when their backs were turned, retook the town and fell upon their rear. Had feudalism been only united, it is certain that no system was so well calculated to arrest the progress of an invasion as this subdivision of the defence; and herein lies the explanation of the incredible facility with which provincial conquests were then lost; for it was not possible at that time, as it is now, to secure the results of a campaign by the centralisation of the military power and by an absolute discipline. If the conquered country was divided into a number of lordships or baronies, which defended themselves each on its own account, much more than to keep the oath of fealty sworn to the suzerain; so the armies were composed of vassals, who were bound to give forty, or sixty, days' service, as the case might be, in the field, and no more; after which term every one might return to his home; and this must be, so long as the suzerain could not have his soldiery on hire. In this respect, from the date of the close of the thirteenth century, the English monarchy had acquired a great superiority over that of the French. The Anglo-Norman feudalism formed a better consolidated mass than the feudalism of France; it had proved this by forcing a reluctant king to grant them their Magna Charta; and as a consequence of this agreement, they were more intimately bound to their suzerain. Their form of government, comparatively liberal, had led the English aristocracy to introduce into their armies troops of foot-soldiers taken from the towns, who were already well disciplined, skilled in the use of the bow, and who decided the fate of the day in nearly all the disastrous[g] battles of the fourteenth century, Crécy, Poitiers, &c. The same feeling of distrust which made the French feudal lord isolate his castle from the town placed under his protection, would not allow him to give arms into the hands of the burghers, or to familiarize them with military exercises; he put his trust in his own men, in the goodness of his horse and of his coat of mail, and, more than all, in his own personal courage; and held in disdain the foot-soldier (*fantassin*) whom he brought into the field only to swell his numbers, not reckoning him on any account at the moment of action. This feeling, which was so fatal to France at the period of her wars with England, and to which may be attributed the loss of many a pitched battle in the course of the fourteenth

[g] Disastrous, that is to say, *for the French*.—TRANSLATOR.

century, in spite of the incontestable superiority of the feudal horsemen (*gendarmerie*) of that country, was essentially favourable to the development of military architecture; and, in point of fact, there is no country of Western Europe where one meets with more numerous, more complete, or finer feudal fortifications, of the date of the thirteenth and fourteenth centuries, than in France[h]. It is in the feudal castles, above all other places, that we must study the military dispositions of those times: for in these they are developed with a profusion of precautions and an abundance of means alike extraordinary.

We have already alluded to the distinction to be drawn between these castles serving as refuges, or as citadels to the garrisons of towns, and the isolated castles which commanded some village, hamlet, or small open town, with which they were only connected by intermediate works. These latter castles were of several kinds: there was the simple keep or donjon, surrounded by a fortified wall, with quarters for soldiers attached; and there was the castle, which occupied a vast space, closed in by strong walls, and containing detached forts

---

h The number of castles which covered the soil of France, more especially on the frontiers of provinces, is incalculable. There was not a village, *bourgade*, or small town, which did not possess at least one, without reckoning the isolated castles, military posts, and towers which at short intervals dotted the courses of rivers, the valleys which were used as passes, and the *marches*. From the earliest period of the feudal organization, the *seigneurs*, the cities, the bishops, and the abbots had to recur on many an occasion to the sovereign authority of the kings of France in order to prohibit the erection of new castles which might be likely to prejudice their interests and "those of the country." (Les Olim.) On the other hand, the king of France, in spite of the resistance of his vassals, authorized by act of parliament the erection of fortresses, in order thereby to lessen the power of his great vassals. "Cùm abbas et conventus Dalonensis associassent dominum regem ad quemdam locum qui dicitur Tauriacus, pro quadam bastida ibidem construenda, et dominus Garnerius de Castro-Novo, miles, et vicecomes Turenne se opponerent, et dicerunt dictam bastidam absque eorum prejudicio non posse fieri. Auditis eorum contradicionibus et racionibus, pronunciatum fuit quod dicta bastida ibidem fieret et remaneret." (Les Olim, edit. of the Ministry of Public Instruction, Philip, III. 1279, vol. ii. p. 147.)

and one or more donjons. Placed on the high-road, or on the bank of some river, castles of this importance could intercept all communications; they formed so many strong places, vast in dimension and of great importance, considered from a military point of view; requiring for their blockade a numerous army, for their subjection a considerable siege train and much time. The Château-Gaillard, in the Andelys, was of this number. Built by Richard Cœur-de-Lion, after this prince had discovered the error he had committed in leaving to Philip-Augustus, by the treaty of Issoudun, the Vexin territory and the town of Gisors, this castle still preserves, in spite of the state of ruin into which it has fallen, the impress of the military genius of the great Anglo-Norman king. Richard, although a bad politician, was a consummate warrior; and it was characteristic of his nature to repair his shortcomings as a statesman by dint of courage and perseverance. In our opinion, the Château-Gaillard des Andelys reveals one portion of the talents of Richard. There is too general a disposition to believe that this illustrious prince was nothing more than a fighter, brave to rashness; but it is not merely by possessing the qualities of a good soldier, however fearless or intrepid, that a monarch acquires so large a place in history. To the men of his time, Richard was a hero whose valour shone conspicuous in a valiant age; but he was also an able captain, an *engineer* full of resources, experienced; a master in the practice of his art, capable of things in advance of his age, and who never allowed himself to be the slave of routine. Thanks to the excellent work of M. A. Deville on the Château-Gaillard[1], everybody can form an exact estimate of the circum-

[1] *Hist. du Château-Gaillard et du siège qu'il soutint contre Philippe-Auguste, en* 1203 *et* 1204 (Hist. of the Chateau-Gaillard and of the siege it sustained against Philip-Augustus in 1203 and 1204), by A. Deville, Rouen, 1849.

stances which regulated the construction of this fortress, the key of Normandy, and a frontier place capable of arresting the invasions of the French king for a considerable space of time. The right bank of the Seine being then in the possession of Philip Augustus as far as the Andelys, a French army could in a single day be conveyed into the very heart of Normandy and menace Rouen. Aware too late of this danger, Richard was anxious to place his continental province beyond its reach. With the sure coup d'œil which belongs to men of genius, he chose the site of the fortress intended to cover the Norman capital, and having once decided upon his plans, he followed out their execution with a tenacity and force of will which bore down every obstacle opposed to his undertaking; so that, in one year, not only was the fortress built, but a complete system, likewise, of defensive works was thrown up, with rare talent, along the banks of the Seine, to the point at which that river covers Rouen. It is rare to find at this period the breadth of view in military dispositions which marks the great soldier; and here it is not merely the isolated defence of a detached post that is in question, but that of the frontiers of a great province. From Bonnières to Gaillon, the Seine flows almost in a straight line towards the north-north-west. Near to Gaillon, it makes a sudden bend towards the north-east, as far as Les Andelys; then turns back upon itself and forms a peninsula which, at its neck, is no more than two thousand six hundred mètres (about $1\frac{3}{5}$ miles) across. The French, by the treaty which followed the conference at Issoudun, possessed, on the left bank, Vernon, Gaillon, Pacy-sur-Eure; and on the right, Gisors, which was one of the strongholds of this part of France. An army composed of corps collected at Evreux, Vernon, and Gisors, and

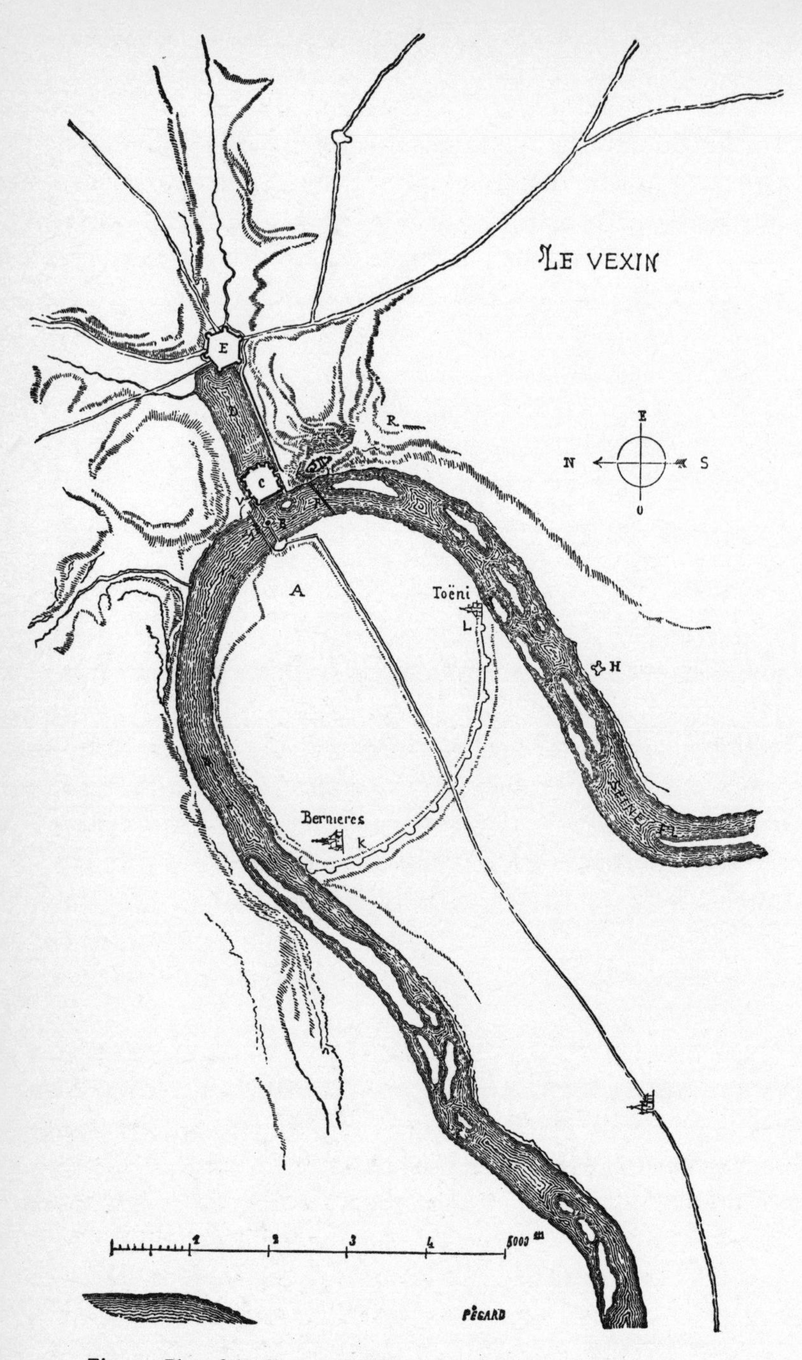

Fig. 27. Plan of the Chateau-Gaillard and its Environs. (See opposite.

thence simultaneously marched upon Rouen, while a flotilla followed in their rear, would be able in two days to invest the place, and have everything they required brought down by the river. By placing a fortress so as to span the river between these two places, Vernon and Gisors, and in such a way as to command the navigation, he prevented the junction of the two corps of invasion, rendered their communication with Paris impossible, and placed them in the awkward predicament of being separately defeated under the walls of Rouen. The position, therefore, was perfectly well chosen. The peninsula which was situate opposite Les Andelys, easily defended across the neck, supported by a fortified place of great strength on the other side of the river, offered every facility for the establishment of a camp which it would be vain to think of forcing. The city of Rouen was covered; nor could the French armies advance against this place without feeling very serious apprehensions respecting the military position they were placing between themselves and France. This short description will serve to shew that Richard was something more and better than a captain full of headlong courage. The manner in which the Anglo-Norman king arranged the plan of his defences, for this strategical position, was as follows, (fig. 27). At the extremity of the peninsula, A, on the side of the right bank, the Seine flows round steep sloping cliffs of great height, which overlook the whole peninsula. On the little island, B, which stands in the centre of the river, Richard erected, firstly, a strong octagonal work with towers, ditches, and palisades

---

A. Head of the Peninsula.
B. A small Island.
C. The Tête-du-pont, or Petit-Andely.
D. The Lake.
E. The Grand-Andely.
F. The Stockade.
H. The Fort, or *Boutavant*.
K, L. Rampart of Circumvallation.
R. The Plateau.

complete[j]; a wooden bridge passing through this fortalice connected the two banks. At the extremity of this bridge, at C, upon the right bank, he built an enceinte, or wide *tête-du-pont*, which was soon filled with habitations, and took the name of Petit-Andely. A lake formed by the retention of the waters of the two streams, at D, completely isolated this tête-du-pont. The Grand-Andely, E, was likewise fortified and surrounded by ditches, which still exist. Upon a promontory of chalk cliff, which rises to a height of more than a hundred yards above the level of the Seine, was planted the principal fortress, advantage being taken of every projection of the rock: towards the south, a tongue of land of no more than a few yards in width served to link this promontory to the surrounding hills. At the base of the escarpment, and enfiladed by the castle, a stockade, F, composed of three rows of piles, was placed across the course of the Seine. This stockade was further protected by palisadoed works erected along the side of the right bank, and by a wall descending from a tower, built half-way up the hill: in addition, a fort was built, at H, on the banks of the Seine, and took the name of *Boutavant*. The peninsula being thus made secure, it was impossible for an army to find an encampment upon ground cut up with ravines and covered with enormous rocks. The small valley lying between the two Andelys, filled by the waters of the stream and commanded by the fortifications of the two *bourgs*, could not be occupied. The single attackable point of the fortress was the tongue of land which connected it with the hills on the south. We will now describe how Richard, who himself presided over the erection of this fortress, and never left the

j The lower portions of this work are still in existence.

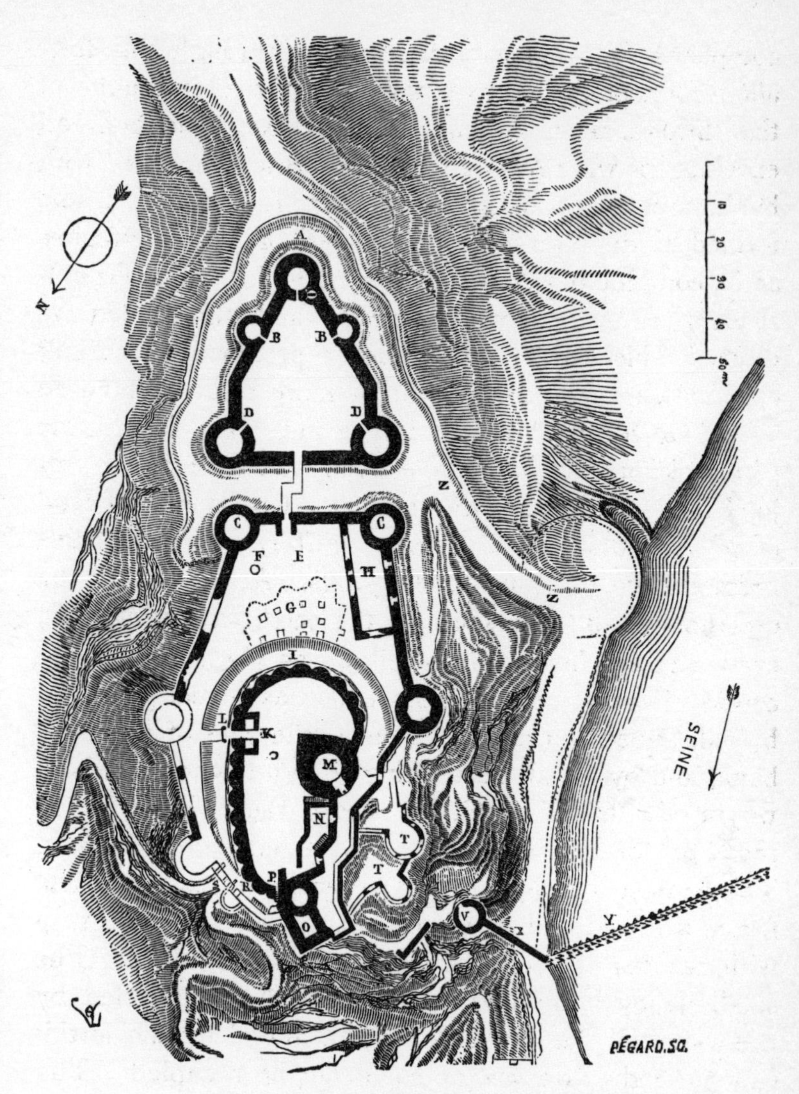

Fig. 28. Ground-plan of the Chateau-Gaillard.

A. High Angle Tower.
B B. Smaller Side Towers.
C C, D D. Corner Towers.
E. Outer Enceinte, or Lower Court.
F. The Well.
G, H. Buildings in the Lower Court.
I. The Moat.
K. Entrance Gate.
L. The Counterscarp.
M. The Keep.
N. The Escarpment.
O. Postern Tower.
P. Postern Gate.
R R. Parapet Walls.
S. Gate from the Escarpment.
T T. Flanking Towers.
V. Outer Tower.
X. Connecting Wall.
Y. The Stockade in the River.
Z Z. The Great Ditches.

workmen, hastening on their labours by his personal direction, established his defences, (fig. 28). At A, opposite the tongue of land which united the site of the castle with the neighbouring height, he had a deep ditch dug in the rock and built a strong and lofty tower, the parapets of which attained to the level of the topmost plateau, and commanded the summit of the hill. This tower was flanked by two others at B B; the curtain-walls, A D, spread out from A and followed the natural slope of the rock; the tower, A, commanded the whole advanced work, A D D. A second ditch, also excavated in the rock, separates this out-work from the body of the place. The two towers, C C, commanded probably the towers D D[k]. The first (or outer) enceinte of the castle, E, contained the stables, domestic offices, and the chapel; this formed the *lower court*. A well existed at F; within and beneath the area of the court, vast cellars were excavated, supported on piers of solid rock, which received their light from the castle-moat, I, and communicated, by means of two tunnels bored in the rock, with the outside. At K is the entrance-gate of the castle, the sill of which is raised more than two yards above the counterscarp of the moat, L. In Richard's days, works erected upon a rock which had been left jutting from the ditch, covered this entrance; a portcullis, overhanging sheds and two small side works, or posts, defended the gate, which was further commanded by the defences of the donjon, M. Soldiers' quarters were arranged on the side of the escarpment, at N, and a strong defence, O, flanked the postern, P, which opened on the parapet-walk, R. It is probable that the gate of the first en-

---

[k] These four towers are now levelled; the plan only can be distinguished, and a few fragments which are still standing.

ceinte opened at S, above the escarpments[1]. On the side of the river, at T, were towers and flank-works stepped along the cliff, in which they were cut, and furnished with parapets: a tower, V, planted against the cliff, and communicating with the body of the place by stairs and galleries excavated in the rocks, was connected with the wall, X, which acted as a barrier across the foot of the escarpment and the river banks, and likewise with the stockade of piles intended to intercept the navigation. The great ditches, Z, descend to the river-side; these were cut in the rock by manual labour. One year had sufficed for Richard to finish all these enormous works and the whole system of defences which was attached to them[m]. "Is she not fair, my daughter of a year?" exclaimed the monarch, when he beheld his great undertaking finished[n]. At the close of the twelfth century the Norman fortifications had nothing in common with the forms adopted in the construction of the Château-Gaillard; we may therefore safely conclude that Richard was alone the author of them, and that he had himself planned and marked out certain arrangements of defence which denote a profound experience in the military art. Had Richard brought back from the East acquirements so far in advance of his age? It is hard to say. Were they the last remains of Roman tradition[o]? Or rather, had this prince, as the result of practical observation,

---

[1] We have left, in line merely, the defences of which only the faintest traces at present remain.

[m] 1196-1197.

[n] "Ecce quam pulchra filia unius anni!" Brompton, *Hist. Angl. Scriptores Antiqui*, col. 1276.

[o] Jean de Marmoutier, a monkish chronicler of the twelfth century, relates that Geoffrey Plantagenet, the grandfather of Richard Cœur-de-Lion, when besieging a certain fortress, used to study the Treatise of Vegetius. (*Hist. du Château-Gaillard*, by A. Deville.)

found in his own genius the ideas of which he then made so remarkable an application?

If we cast our eyes on the plan (fig. 28), we must be struck by the curious configuration of the elliptical enceinte of the internal castle; it consists of a succession of segments of a circle, the chords measuring about three yards, which are separated from each other by portions of the curtain-wall only a yard in length. In plan, each of these segments gives the following figure (fig. 29), which forms a

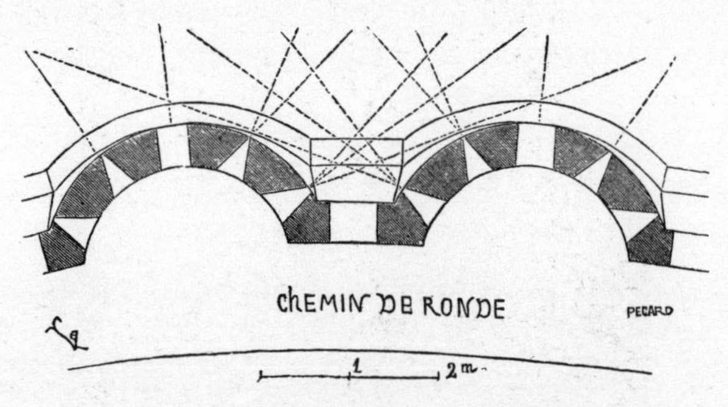

Fig. 29. Chateau-Gaillard—Plan of Segments.

continuous series of flank-works of great strength compared with the offensive engines of that period, as is shewn by the dotted lines. In elevation this wall, whose base rests upon the hewn rock, presents a formidable appearance (fig. 30). There is no opening of any kind in the lower portion, the whole of the defences being arranged at the summit [p]. The donjon is no less interesting as a study: it consists (fig. 31) of a series of reversed pyramids, with their bases round the summit, attached to the flanks of the tower. These pyramids must have had, springing from their reversed bases,

---

[p] The walls have now fallen to the level of the point O: it is probable that hoards or brattishes were attached to the anterior portion of the segments in time of siege, as we have shewn at B. But this is mere conjecture.

Fig. 30. Chateau-Gaillard—View of part of the Wall. O. (See opposite.)

arches to form a machicolation[q] for close defence, and have supported an embattled parapet for distant defence. Dying away upon the lower splay of the tower-wall, pyramids so placed offered projections very judiciously combined for commanding the base of the donjon, while at the same time they consolidated the whole work. On the side of the castle this donjon presented a projecting angle (see the plan, fig. 28), which increased the resisting force of the masonry at the sole point where it was possible to have sapped it.

In all these works no sculpture is to be seen, or mouldings of any kind; everything has been sacrificed to the defence: the masonry is good, and composed of a rubble of silex bedded in excellent mortar and revetted (or faced) with carefully executed face-work in small courses, here and there having alternate courses of red and white stone.

During the life-time of Richard, Philip Augustus, notwithstanding his well-earned reputation as a taker of fortresses, dared not venture to lay siege to Château-Gaillard; but after the death of this prince, and when Normandy had fallen into the hands of John Lack-land, the French king resolved to seize on this military position, the possession of which would open the gates of Rouen. The siege of this place, described to its smallest details by the king's chaplain, William the Breton, an eye-witness, was one of the greatest feats of arms of this prince's reign; and if Richard displayed a talent every-way remarkable in the strategetical arrangements of the place and its dependencies, Philip Augustus carried out his enterprise like a consummate master in the art of war.

---

[q] These crowning members no longer exist: the structure is level with the point O. The view we give is taken from the side of the postern of the donjon, on the north, which is placed on the first floor. We have supposed the building, N, as removed, in order to allow the staircase which led to this postern to be seen.

Fig. 31.

Château-Gaillard was defended by Roger de Lacy, constable of Chester; and with him a great number of knights of renown were enclosed in the fortress. The French army invested the peninsula of Bernières (see fig. 27), resting their left on this village, their right at Toëni. The wooden bridge, which connected the small peninsula with the fortalice situate on the Isle of Andely, was immediately destroyed by the Anglo-Normans. Philip Augustus caused, first of all (and it was well for him he did so), a ditch to be excavated from one village to the other, and erected a rampart of circumvallation, K L. In order to be able to bring up the necessary boats to form a bridge opposite the Lesser Andely for passing to the right bank of the river, he had the stockade cut by some bold swimmers, who effected a breach in it whilst a false attack was being made on the fortalice (châtelet). This breach once effected,—

"The king," says William the Breton, "had conveyed from divers ports on the Seine a great quantity of those flat boats which serve habitually for the passing over of men, beasts of burden, and carts (called *bacs*), and fastening these together, side by side from one bank to the other, he laid upon them a good floor of planks. The boats which carried these planks were fastened to strong piles driven down here and there in the bed of the river, and were armed with turrets at certain intervals. Four larger boats, M, were attached to the central portion of the bridge, which bridge rested upon the lower point of the Isle of Andely; and upon these boats two great turrets, sheathed with iron, were erected for the purpose of battering the fortalice."

This being done, the French army passed over to the right bank and encamped at V, under the walls of the Lesser Andely. Meanwhile, John attempted to relieve the place: to this end he despatched an army intended to be thrown, during the night, upon the rear of the French at the neck of the peninsula, whilst a flotilla starting from Rouen was at the same time to attack the bridge of boats; but the two attacks were not made simultaneously; the line of circumvallation arrested the attack by land, and gave the French camp time for preparation, whilst the flotilla, which arrived too late on the scene of action, was driven back with considerable loss. The fortalice was soon taken, as likewise the Lesser Andely, and occupied by French garrisons. Philip Augustus was then in a position to lay siege to Château-Gaillard; he pitched his camp on the plateau, R, opposite the tongue of land which connects the castle with the mountain. But winter was drawing near, and the king hoped to take the place by famine. Invested on all sides, the garrison retired within the triple enceinte of the fortress; lines of contravallation and of circumvallation, still visible, were marked out and furnished with seven wooden turrets at regular intervals. During the whole of the winter of 1203-4 the French army remained within the lines. In the month of February, 1204, Philip Augustus, who then knew that the garrison were provided with provisions sufficient to last them over the year, decided on undertaking a siege in form. Opposite the angle tower, A (fig. 28), he had the ground of the tongue of land, which this tower commanded, levelled, and upon the site thus prepared he established covered galleries (*cats*), and a turret which operated against the tower; the ditch was filled, and pioneers were attached to the base of the tower, A, above the rocky escarpment; in a short time the tower fell upon its burnt shoring timbers, the garrison abandoned the advanced work, and the first enceinte of the castle fell into the hands of the king by a surprise, which was effected in this wise. Five

French squires, whose names William de Breton has preserved, obtained an entrance into the building, H, through a window but slightly raised above the moat, and by their loud shouts, suddenly raised, made the garrison believe that a numerous body of troops had invaded the first enceinte; whereupon the besieged themselves fired the buildings of the lower court-yard and retired into the castle. After incredible efforts, Philip Augustus succeeded in placing in battery, opposite the gate of the castle, K, a catapult, and in attaching his pioneers to the work which defended this gate, by advancing a cat upon the position. In a little time the gate was shattered and a portion of the masonry fell. The French precipitated themselves upon the breach with such impetuosity, that the garrison, then reduced to one hundred and eighty men, could not force a passage to the postern of the donjon, and being surrounded, they were obliged to lay down their arms. This happened on March 6, 1204. The first operations of the attack on the fortalice and the passage of the Seine took place in the preceding month of August. It is evident that under another prince than John, Château-Gaillard would have held out much longer; for the besieging army, harassed by an enemy from without, would not have been able to proceed so methodically and with such united action. The journal of this siege places on evidence a fact which is curious in reference to the history of fortification. Château-Gaillard, in spite of its situation, in spite of the great skill displayed by Richard in the details of its defence, is too restricted in its dimensions. Already even, for that period, the arrangements of defence which were accumulated upon a given point, instead of supporting each other, were mutually injurious; the means of attack, as they became more energetic and powerful, called for a more extended line of defences. We shall see presently how, during the thirteenth century, engineers simplified their fortifications and subjected them to methods of more regularity and of greater breadth.

The castle of Montargis, the construction of which dated from the thirteenth century, and a plan of which we subjoin (fig. 32), was likewise a place of sufficient

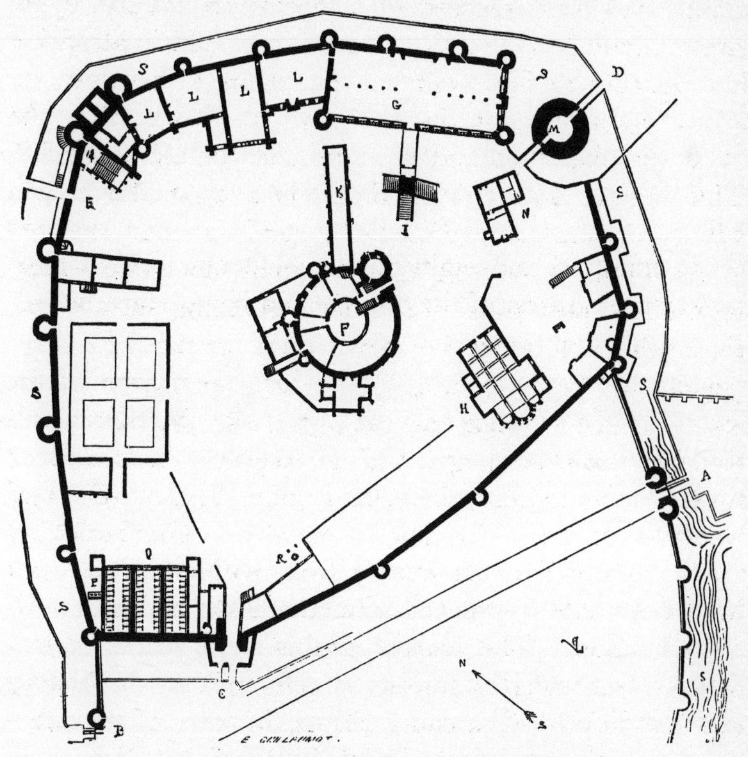

Fig. 32. Plan of the Castle of Montargis.

A & B. The Outer Gateways.
C. The Inner Gateway.
D. Another Entrance.
E. The Postern.
F. The Keep.
G. The Great Hall.
H. The Chapel.
I. The Staircase.
K. The Gallery.
L L L. The Barracks.
M. A Gateway Tower.
N. A Guard Tower.
O. The Stables and Offices.
S S S. The Moats.

strength to call for a regular siege. It commanded the high road from Paris to Orleans which passed through the fortified gates, A and B. Moats, S S, surrounded the external and internal defences. The road was exposed to a flank fire from a front flanked by towers, and communicated with the castle by means of a gate, C (see fig. 61, for a bird's-eye view of this entrance); another gateway, passing through a massive detached tower, was of very difficult access. As for the internal arrangements of the castle, they are of great interest, and shew clearly the means of defence then in the hands of a garrison. The towers are of great projection beyond the line of the curtains, with a view to obtain a good flank fire; on the north (a salient point, and weak in consequence) was erected an important work composed of two massive walls, one behind the other, connected by other return walls, which latter were flanked by two towers of a greater diameter than the others. At G was the great hall, two stories high, in which the whole of the garrison could be called together to receive orders, and from which they could rapidly be directed simultaneously upon every point of the enceinte, by means of a staircase of three flights, each in a different direction, I. The connection between this staircase and the great hall could be cut off, and the great hall be made to serve for a place of retreat, if the enceinte were forced. The massive donjon, F, several stories in height, with a circular tower in the centre, communicated with the great hall on the level of the first story, by means of a gallery, K, which in the same manner could be isolated at its extremity. This donjon commanded the whole of the enceinte and the buildings attached. The garrison was quartered in the buildings, L, on the side where the enceinte was most easily accessible. At O were the stables, the bakehouse, stores; at H the chapel, and at N a *poste*, or guard-house, close to the entrance, D.

The small buildings which surrounded the donjon were of a date posterior to its erection. The postern, E, gave access to extensive gardens, which were themselves surrounded by a fortified wall.

The donjon was to the castle, during the feudal period, what the castle was to the town,—its last retreat, the last means of resistance: and we find it, therefore, constructed with the utmost care and furnished with every means of defence then in use. During the Romanesque period, the donjon is, as a general rule, built upon a square plan, and strengthened by buttresses of rectangular or semicircular form, which had the advantage of flanking the walls by means of battlements placed at

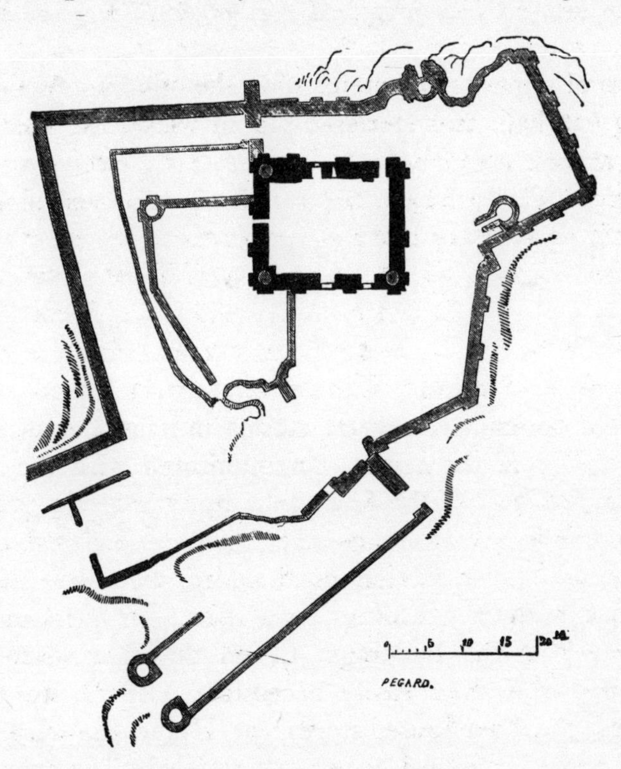

Fig. 33. Plan of the Castle of Chauvigny.

their summits. Such are the donjons of the castles of Langeais, of Loches, of Beaugency-sur-Loire, and of Chauvigny (the plan of which last we append, fig. 33 [r]), of Montrichard, of Domfront, of Nogent-le-Rotrou, of Falaise, &c. Their stories were vaulted over, or separated by timber floors resting upon a row of detached piers; the windows which gave light to these halls were few, and they were frequently furnished with chimneys, an oven, and wells on the ground-floor. They were so contrived as to be built upon the most elevated point of the plateau on which the castle was placed, or on *mottes*, or mounds, made artificially. A wall of counter-guard (or *chemise*) of some height protected their base, and access to the interior could only be obtained by means of a narrow postern raised several yards above the ground-level, and by means of wooden stairs or flying bridges communicating with the parapet of the chemise. As early as at this period the elevation of the donjons was considerable, being from thirty to fifty yards, in order to command not only the exterior defences of the castle, but even the parts outside. There exist no longer, as far as we know, any donjons built from the tenth to the close of the eleventh century the upper defences of which have been preserved; and we are therefore not able to say whether their battlements were furnished with hoards in time of war, or whether they were crowned with platforms or with high-pitched roofs.

However, as in the upper portions of the castle of Carcassonne which are preserved, and which date from the end of the eleventh or beginning of the twelfth century, we have discovered traces perfectly visible of

---

[r] This plan gives the present state of the castle at the height of the first story. The erection of this donjon dates from the eleventh century.

these timber hoards, it is highly probable that the square towers of the west, north, and centre of France were defended in a similar manner. Towards the middle of the twelfth century the square form was abandoned, in donjons as well as in towers; because the salient angles of towers on a square plan, not being capable of a good defence, allowed the besiegers to undermine these angles and thus destroy the whole work. The keep of Etampes offers a peculiar arrangement, and one which shews the efforts which were made in the twelfth century to make these important defences at once feudal residences and well-guarded works. We subjoin (fig. 34) the plan of

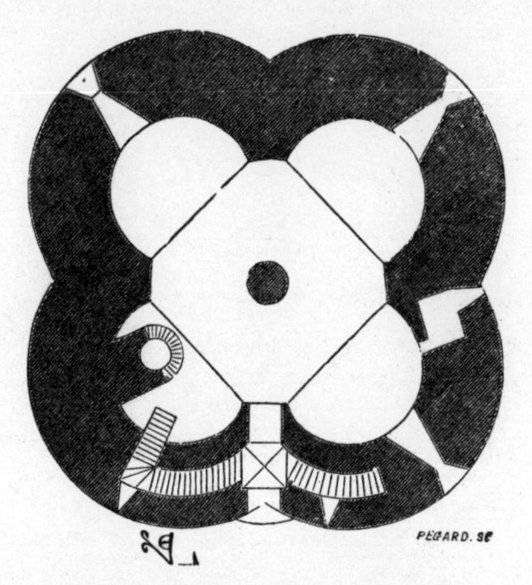

Fig. 34. Plan of the Keep of Etampes.

the ground-floor of the keep[s]. The keep of Provins, called Cæsar's Tower, built in the twelfth century, is still more interesting as a study; it is a complete polygonal (octagonal) fort, flanked by four towers engaged at their base, but which detached themselves from the body of the structure in their upper portions and

---

[s] This plan is drawn to a scale of $\frac{1}{400}$.

thus commanded the ground all round. This work could be manned by a great number of defenders, on account of the receding plans of its several stories and of the flanking position of the turrets. We give (fig. 35) the

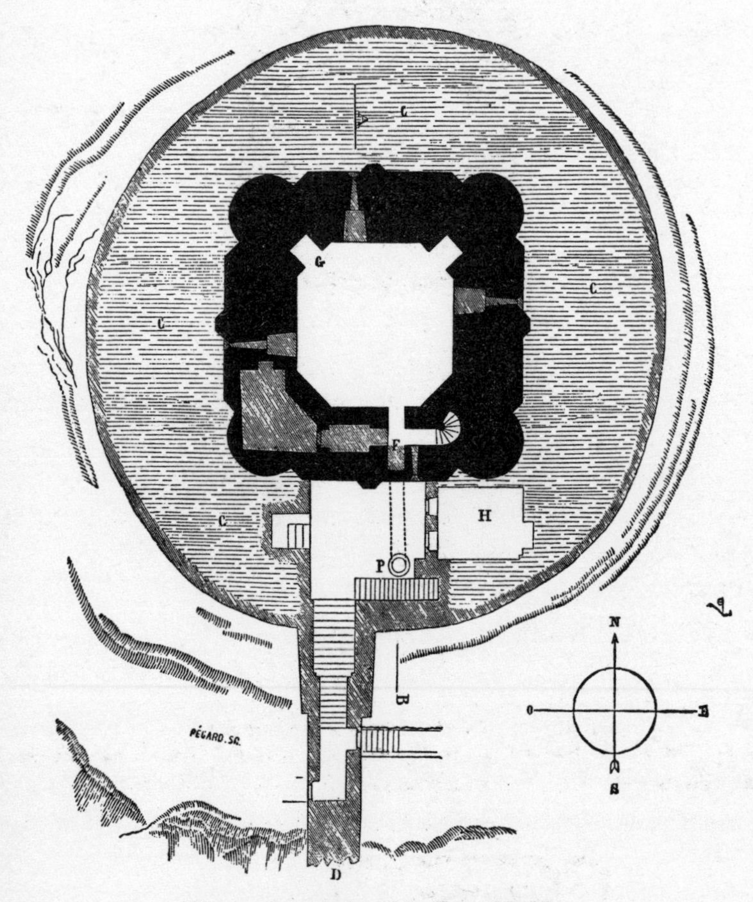

Fig. 35. Ground-plan of the Keep of Provins.

A B. Line of the Section, fig. 39.
C. The Outer Platform.
D. Passage to Town Wall.
F. The Doorway.
G. The Oven.
H. The Chapel.
I I. The Posterns.
P. The Masked Well.

plan of the ground-floor of this donjon; (fig. 36) the plan of the first story; (fig. 37) the plan of the third story

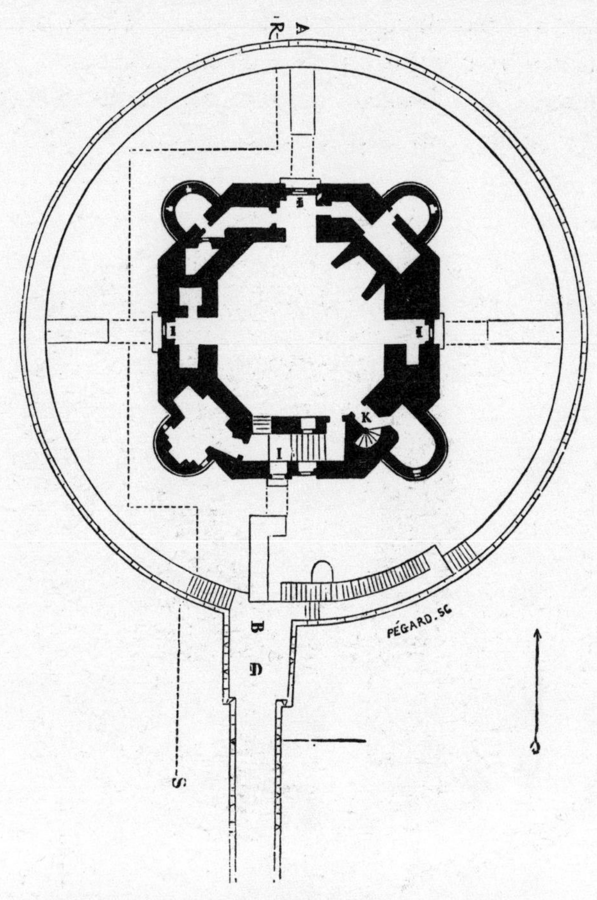

Fig. 36. Plan of the First Story of the Keep of Provins.

A B. Line of the Section, fig. 39.
C. Outer Platform.
D. The Passage.
I I I. The Posterns.
K. The Staircase.
R S. Line of the Elevation, fig. 38.

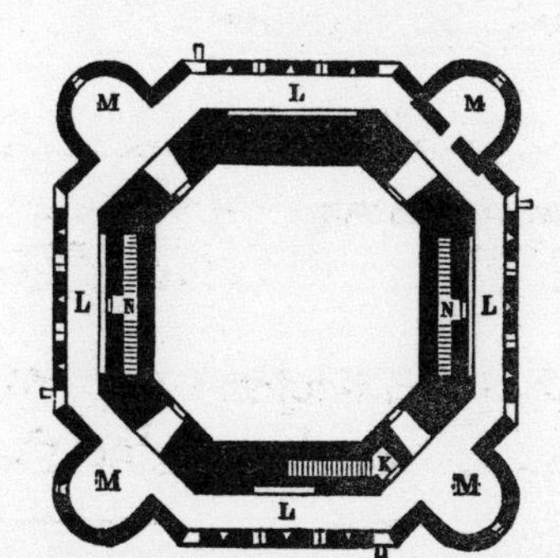

Fig. 37. Plan of Third Story.

K. Staircase.
L L L L. The Allure.
M M M M. Bartizans.
N N. Steps to Upper Platform.

and of the first circulating gallery (*tour-de-ronde*) of defence; (fig. 38) the elevation of the western side, and (fig. 39) the section on the line A B[t]. The platform, C (fig. 35), which surrounds the keep of Provins dates

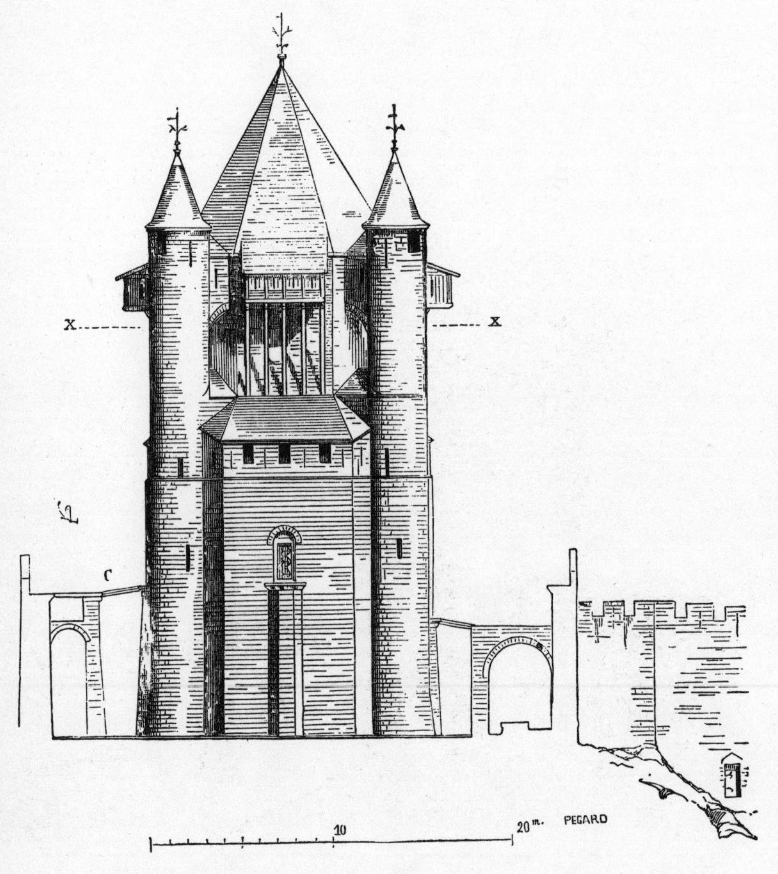

Fig. 38. Elevation of the Keep of Provins, on the line R S on the Plans.

X X. Line of the Present Remains[u]. **C. The Platform added in the 15th century.**

---

[t] To the same scale of $\frac{1}{400}$, full size.

[u] All above this line has been destroyed, and is here restored in the drawing only, from other examples and illuminated MSS.

from the fifteenth century, and was erected by the English to receive cannon; it took the place of a wall of counter-guard belonging to a much more ancient date. The wall, D, was prolonged to reach the Paris gate, and established a communication between the platform, C, or

Fig. 39. Section of the Keep of Provins, on the line A B on the Plans.

X X. Present Level of the Building.

the wall of counter-guard, and the parapets of the town walls. Anciently, access was obtained to the hall on the first story of the donjon, from the parapets of the wall of counter-guard, by means of four posterns, I, (fig. 38) communicating with as many flying bridges. It was necessary to descend from the first story to the ground-floor, which had no communication with the outside; through the doorway, F (fig. 35), you arrived, by a descending flight of steps, at the masked well, P (see section, fig. 39). A dungeon which is traditionally pointed out as the place of imprisonment of John the Good, Duke of Brittany, constitutes, with the great central hall, the ground-floor: at G, an oven had been set up in the fifteenth century; a small chapel was placed at H. The first story is composed of several small chambers, intended for the quarters of the persons in command. By means of the four posterns, I, I, I, I, and the drawbridges, the garrison easily spread themselves along the platform or allure of the original enclosure (*chemise*), which we have marked as restored in the plan (fig. 36), and so passed out to the prolonged wall, D, communicating with the exterior. By the small spiral stair, K, access was obtained to the embattled parapets, L (fig. 37), and the bartizans, M. Finally, by the steps, N, the second story was reached, the defences of which are partially destroyed. The ancient buildings in their present state reach no higher than the level X X (figs. 38 and 39). There is no doubt that the upper portion of this tower was defended with great care, a fact which is proved by the arrangement of the angle-turrets. We have attempted, in the elevation and section which we give, to restore this upper portion, in strict conformity with the defences which exist of this period, and with the indications to be found in manuscripts anterior to the thirteenth century; which indications, however, it must be allowed, are extremely insufficient. The position of the timber hoards of the four upper faces appear to us as placed beyond doubt, as there could be no other explanation of the recess left over the continuous gallery (fig. 39), and which appears intended to receive the feet of the great struts of the hoards; these hoards being of sufficient projection to form a machicolation beyond the line of the first story parapets. The hoards thus placed flank the turrets, and these, in their turn, flank the faces of the tower.

But, in the thirteenth century, they appear to have abandoned all square and angular forms, in setting out their donjons, in order definitively to adopt the circular plan. About the close of the twelfth, or the beginning of the thirteenth century, the donjon of Châteaudun, and that of the Louvre, were erected on a circular plan. Somewhere about the year 1220, Enguarrand III. de Coucy erected the admirable donjon which is still in existence. We shall close this part of our subject by giving a detailed account of this donjon, as being the largest, the most complete, as well as that in which the system of defence is the strongest, and, at the same time, the most easily explained, of all those known to us. We have given (fig. 26) the site of the castle of Coucy, and its position in relation to the town. From the place d'armes, or lower court-yard, in which the domestic offices were placed, an entrance is obtained into the castle over a bridge, A (fig. 40), flanked by two guard-houses. This bridge could easily be cut off in time of war, its causeway resting merely upon detached piers. The great keep, B, and its chemise, commanded at once the lower court-yard and the moat, the back of the surrounding curtains, and the whole of the castle. The towers, H, H, H, H, belong to the same date as the donjon, as likewise the chapel, D. The upper portions of the entrance-gateway, and the great halls, E, F, were

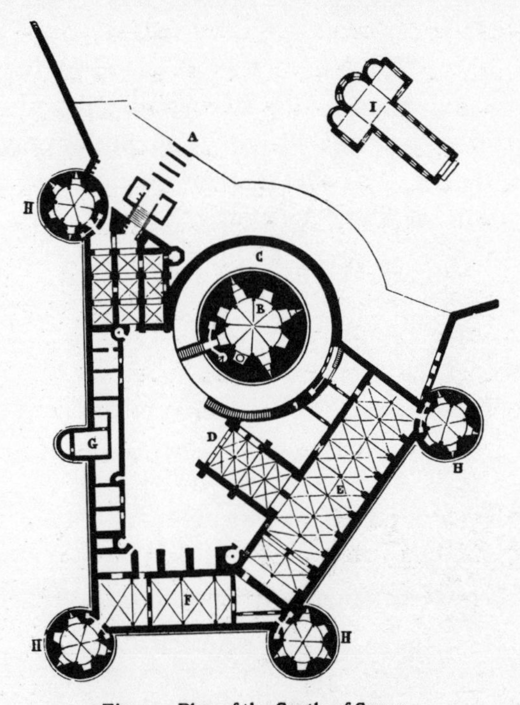

Fig. 40. Plan of the Castle of Coucy.

A. The Bridge. B. The Keep. C. The Ditch. D. The Chapel. E & F. The Great Halls. G. A Bastion. H H H H. Towers.

rebuilt at the commencement of the fifteenth century. If the castle were taken, the garrison retired within the keep, to which the only mode of ingress was through a single doorway, provided with a drawbridge: a ditch, C, isolated the donjon from its chemise. The entrance-passage, A (see the plan of the ground-floor, fig. 41), was defended by a portcullis, two doors, and an iron railing, or *grille;* on the right, a spacious staircase leads to the upper stories; on the left, a corridor which gives access to the necessaries, B. This ground-floor was dimly lighted by a few narrow windows, placed at a great height above the ground (see the section, fig. 44, upon the lines O P of the plans), and probably by the central eye or opening of the vault, which in all probability was

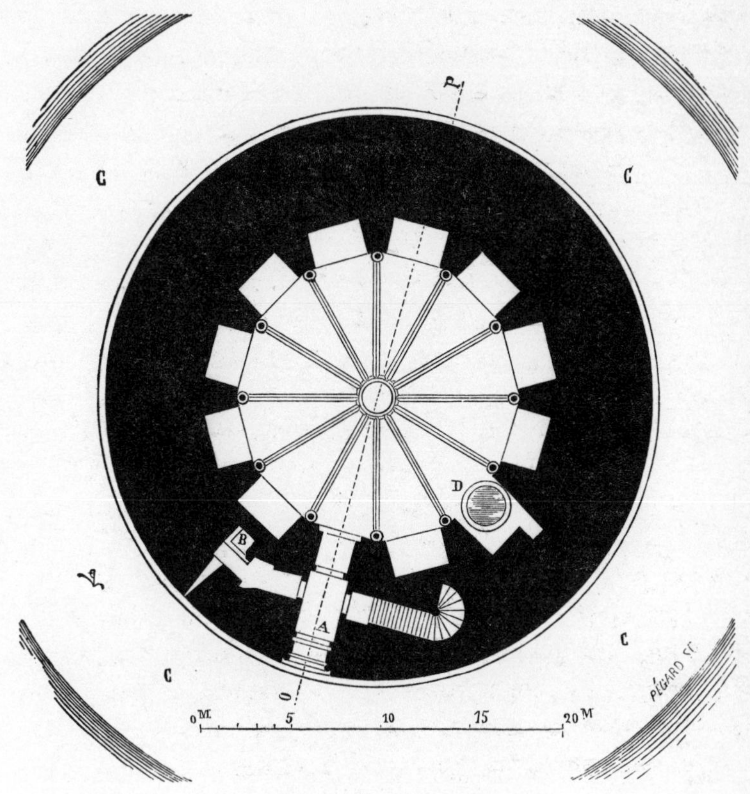

Fig. 41. Ground-plan of the Keep of Coucy.

A. The Entrance. B. The Garderobes. C. The Ditch. D. The Well.
O P. Line of the Section, fig. 44.

repeated at each story up to the top, to facilitate the communication of orders and the hoisting of projectiles, and to admit air into the building. A wide and very deep well was sunk at D, in one of the eleven recesses which surround the hall; in the second bay beyond this well is a fireplace. The vaulting, which is now destroyed, but the springing-courses of which still remain, rested upon sculptured capitals of fine design and on corbels representing figures in a bending position. The first story offers a plan similar to that of the ground-floor (fig. 42). Beneath the recessed sill of one of the windows a closet is constructed, giving access to a passage made in the thickness of the wall, and communicating with a sally-port, D, which, by means of a drawbridge, enables the garrison to reach the parapets of the

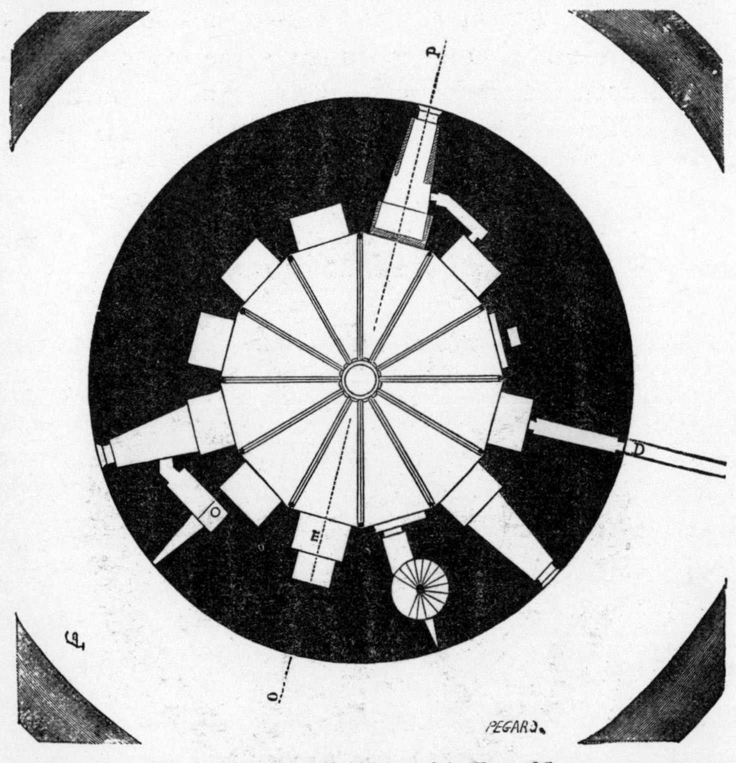

Fig. 42. Plan of the First Story of the Keep of Coucy.
D. The Sally-port. E. The Fireplace. O P. Lines of the Section, fig. 44.

chemise. The fireplace of this story is at E. The second story (fig. 43) offers an admirable arrangement; it consists of a great hall, surrounded by a gallery, the floor of which is raised some ten feet above the pavement of the hall, whilst wooden balconies placed at G, the marks of which are everywhere apparent, enabled those in the hall to advance as far as the inner circumference formed by the upper extremities of the piers. Here it was the whole garrison was assembled, when general orders had to be given out. From twelve to fifteen hundred men could, by means of the gallery and balconies above mentioned, be easily collected in this immense rotunda, and whatever was said at the centre could be heard by all. We know of nothing, either in the monuments of Roman antiquity or in our modern edifices, which possesses an appearance at once so strikingly grand, and so

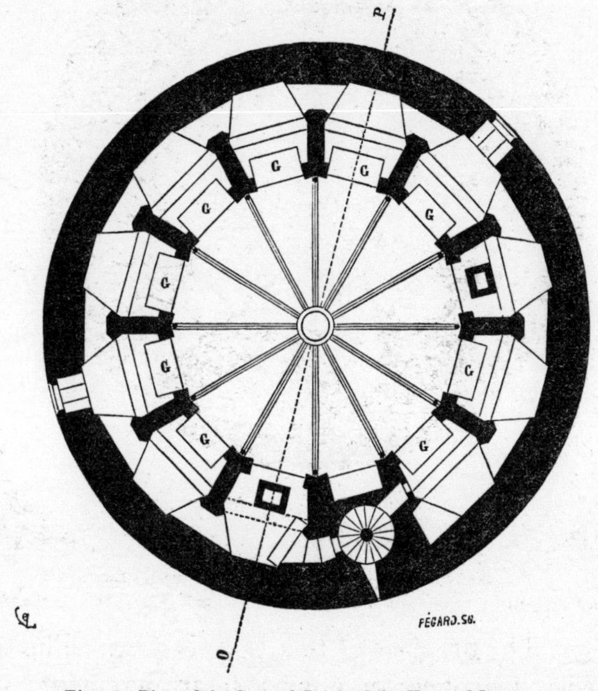

Fig. 43. Plan of the Second Story of the Keep of Coucy.

GGGG. Wooden Galleries. O P. Line of the Section, fig. 44.

stamped with the impress of power, as this beautiful structure; of which, indeed, our section (fig. 44) can convey but a very feeble idea. Still ascending the spiral staircase, we reach the battlemented story (fig. 43). A flagged or leaded covering protected the vaulting and formed an inclined platform, round which a wide walk or allure allowed of a free passage and access to the parapet. The channels for the flow of surface water, which are carefully constructed in the haunches of the vaults over the gallery, prove beyond a doubt that this story was always uncovered; but in time of war a line

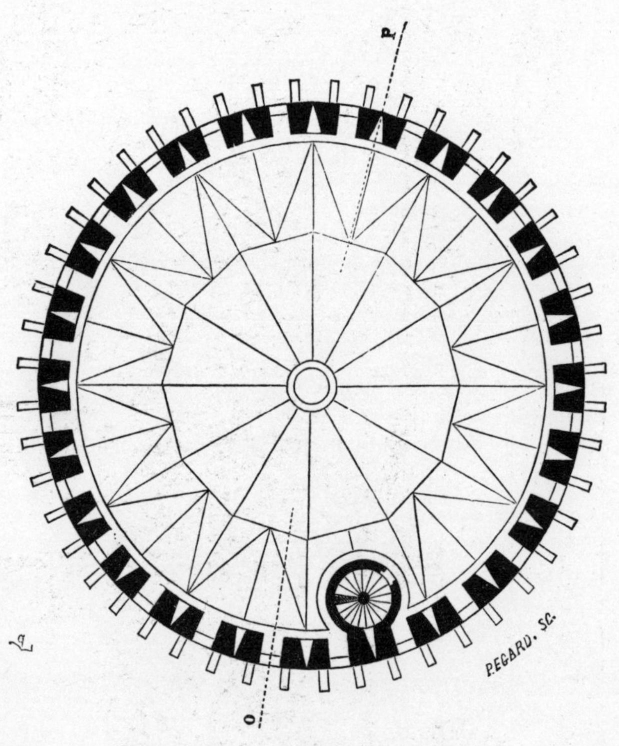

Fig. 44. Plan of the Platform (on the Roof, the Allure behind the Parapet, and of the Battlements) of the Keep of Coucy.

of two-storied hoarding was placed upon stone corbels, built into the thickness of the wall below the battlements (fig. 44). We here see the first appearance of the transition from timber hoarding to stone machico-

Fig. 45. Section of the Keep of Coucy, on the line O P of the Plans.

lation. For a work, indeed, so powerfully conceived and executed as this was, wooden hoarding resting upon overhanging beams must have appeared a defence not sufficiently durable. This system, of timber hoarding resting upon stone corbels, is applied not only to the keep at Coucy, but likewise to the towers of the castle. Nor is it the defensive dispositions at Coucy alone which are calculated to attract the attention of the architect and antiquary; the keep retains, as we have already mentioned, fragments of sculpture of the highest beauty; and everywhere may be found the traces of coloured ornament, exceedingly simple but executed in a fine style. There are still several curious facts in connection with the construction of this immense fortress which require notice. Everything therein which may be classed as a matter of general use (such as the seats, the steps of stairs, the sills of the upper windows,) appear as if intended for a race larger than man: the benches are 2 feet high, the risers of the steps from 12 to 16 inches; the sills of the windows are 3 feet 6 inches high. The materials built into the work are of enormous dimensions; we find lintels of doors not less than two cubic yards, and courses of stone 27½ inches in height.

The following appears to have been the plan followed in the erection of the donjon at Coucy: the construction was carried on spirally from the base to the summit, by means of a scaffolding which was fixed as the works proceeded; this scaffolding, erected outside of the external face of the wall, formed an inclined tramway, by means of which the largest stones could be wheeled up to the summit. The square holes of the transverse beams of this scaffolding, and of the braces which kept them in position, are still visible, very regularly disposed on the circumference of the enormous cylinder. It was impossible to adopt a course at once more simple or more ingenious for building rapidly, and without useless expense, a tower of such dimensions; a tower which is not less than 100 feet in external diameter, and 200 feet in height from the bed of the moat to the bottom of the water-table which surmounted the enriched cornice at the summit. At the present day the vaults of the two stories have fallen in, and the upper water-table, or coping, above referred to, as well as the four pinnacles which crowned it, no longer exist. This crowning member is mentioned by Ducerceau in his book, "On the most excellent Buildings in France" (*Des plus excellents bâtiments de France*); we meet with some fragments of it on the upper parapets and at the bottom of the moat, but of the pinnacles we have not been able to discover any remains; excavations made in the moat would probably lead to their being partly recovered[1]. The whole of the masonry was chain-bonded by means of wooden wall-plates, from seven to eleven inches square, built into the thickness of the walls, according to the mode still in use in the twelfth century. Above the vaulting of the first story this timber was linked to a system of radiating bars, also of wood.

It would seem as if this keep had been built for a race of giants, and the appearance of the structure is in harmony with the power displayed in its execution: fitting dwelling-place for that Enguerrand III. de Coucy who is indeed the greatest figure of the feudal age. We must bear in mind that this heroic personage, after having

[1] It is to be desired that the Government would authorize excavations to be made in the castle of Coucy; as there is there a mine of precious information bearing upon the history of architecture, as applied to the military art of the Middle Ages. We excavate at Nineveh, but leave buried at a few leagues from Paris traces, still vital in their interest, of the history of France.

ravaged the dioceses of Rheims and Laon; after King Philip Augustus had made to the Chapter of Rheims, who complained to him of his acts of violence and of the ravages he had committed on their lands, the celebrated answer, "I can do no more for you than *pray* the Sire de Coucy to leave you unmolested,"—aspired, under the monarchy of Louis XI., to the throne of France. He was lord of Montmirail, of Oisy, of Crèvecœur, of la Ferté-Aucoul; possessed the lands of Condé in Brie; was Count of Roucy, Viscount of Meaux, and Castellain of Cambrai. Fifty knights were always round him, independently of the vassals who owed him military service. Fifty knights, with their following, formed a guard of about five hundred men. In the thirteenth century such a position, and a castle like that of Coucy, placed a vassal of the king of France on a footing of equality with his suzerain. But although it was given to only a small number of the vassals of the crown of France to take so high a place or acquire such immense riches, and an influence so considerable, all of them in varying degrees wished to preserve their independence, and to keep up a perpetual struggle with a society already aspiring to monarchical unity; all of them erected castles: there was not the smallest of the seigneurs but had his nest, his barred refuge, and his men: and in time of war he sided with such or such party; now for his feudal suzerain the king, at another time for the foreigner; according as he thought he could obtain honour, or profit, or, it might be, the satisfaction of a personal revenge.

How poor soever the castle might be, advantage was always taken, as far as possible, of the natural escarpments of the ground, when it was being erected; for thereby it was placed beyond the reach of engines of war, of the sap and of the mine. As the attack was never made except close to the walls, and as the catapults and other projectiles of that nature could not hurl their projectiles to a very great height, there was a great advantage in commanding the assailants either by a natural escarpment of crag, or by structures of a great elevation; whilst means for resisting the external enemy, on the level of the plane of attack, were prepared in the lower portions of the towers and curtain-walls. While, under the influence of the monarchy, feudalism was undergoing a process of subdivision, it made up for its decreasing resources by calling to its aid the most active means of defence; it exerted all its ingenuity in placing its castle in a position to resist the most formidable attacks; it multiplied the obstacles round its places of refuge; hither tended its constant anxiety, this was the end and aim of all its sacrifices, and the best use to which its revenues and the wealth derived from the deeds of prowess of its members could be put. In this way also it served to give a powerful impulse to the progress of the art of fortification.

We have already seen that the towers of the ancient Romanesque period had their lower portions unexcavated, and the curtain-walls were revetted terraces of earthwork. From the beginning of the twelfth century the inconvenience attending this mode of construction had been felt, as it gave the besieged merely the tops of the towers and curtain-walls for his defence, and left all the basement and foundations open to the miners and pioneers of the enemy; thus the latter could place shoring timbers under the foundations and bring down great lengths of wall by setting fire to these props, or sink the gallery of a mine under the foundations and earth-work, and thus obtain an opening into the interior of the works.

In order to meet these sources of danger, the military engineers of those times constructed lower stories in their towers, beginning at the bed of the ditch, the level of the water, or the upper surface of the rock escarpments; these stories were provided with loop-holes, radiating in the manner indicated at fig. 46, so as to be able to direct a fire from every point of the circumference, as far as this was practicable. The same arrangement was adopted in the curtain-walls, especially wherever they served as the outer walls of buildings divided into stories, which in castles was almost always the case. The pioneers had thus increased difficulty in arriving at the foot of the walls, for they were obliged to protect themselves not only against projectiles flung down from the top, but likewise against arrows fired obliquely and horizontally through the loop-holes; if they succeeded in effecting a breach at the base of the wall of the tower, they were sure to find themselves opposed by a force of the besieged, who, made aware by the noise of the sap of what was going on, were enabled to throw up a palisade, or a second wall behind the breach, and thus render their labour vain. So that when the assailants, by means of their engines, had dismounted the hoards, dismantled the battlements, filled up the moats; and when with his companies of archers and cross-bows, whose fire swept the summit of their ramparts, he had at last made it possible for his pioneers to get to work; these latter, unless they were both very numerous and very bold, and unless they could throw up trenches of great width and bring down an entire work at a time, found, behind the opening they had effected, an enemy awaiting them in the lower rooms of the works, on the ground level. Should the assailants even succeed in forcing their way into these works by killing

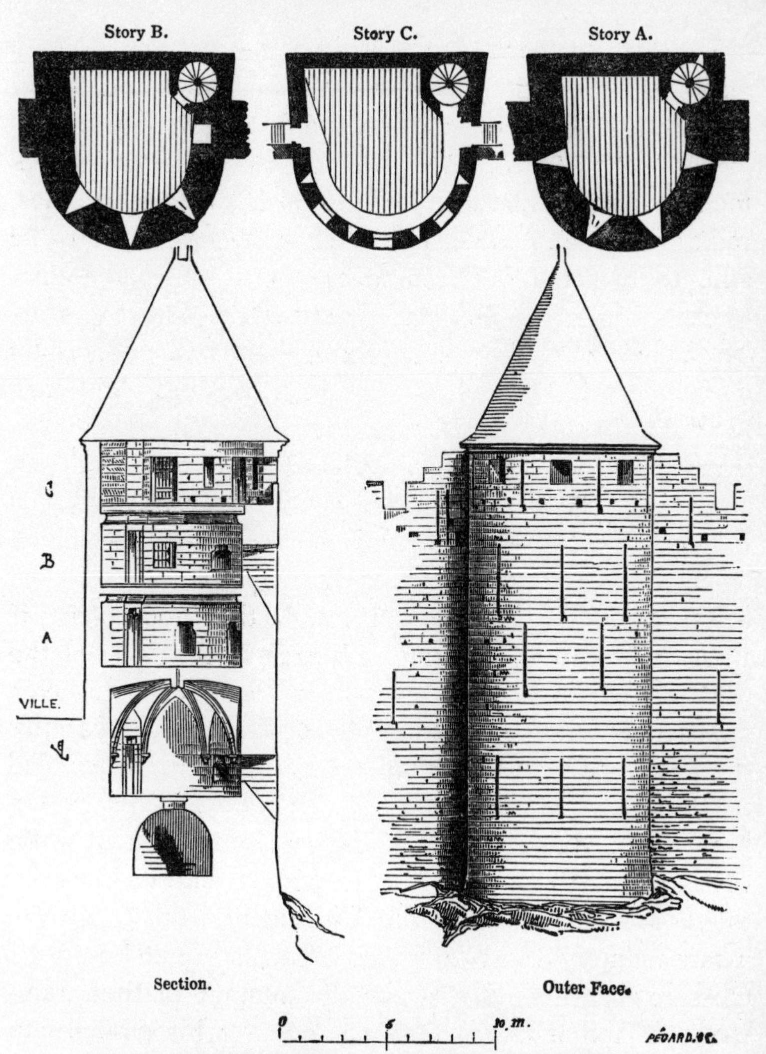

Fig. 46. Elevation, Section, and Plans of a Tower at Carcassonne.
A A. Section and Plan of the First Story. B B. Section and Plan of the Second Story.
C C. Section and Plan of the Third Story.

the defenders, they would still have to gain access to the upper stories, up narrow staircases easily barricaded and guarded by doors and iron gratings (*grilles*).

It is worth observation that the out-works and the towers of the lists were pierced with loop-holes of a form permitting the besieged to employ a horizontal fire, in order to defend the approaches at a great distance, whilst the loop-holes of the towers and curtains of the second enceinte were made to facilitate a plunging or vertical fire. These openings, however, which on the outside were no more than some four inches in width, widening to a yard or a yard and a-half inside, served rather for reconnoitring the enemy's movements, and for letting air and light into the interior of the apartments of the towers, than for defence; the angle at which they commanded the outside was too acute, especially when the walls were of a great thickness, to allow of the arrows, bolts, or quarrels fired through these narrow slits to do any serious damage to the assailants. The real defences of the tower were placed at the top of the works. There, in times of peace, and when the hoards were not mounted, the parapet wall, of a thickness varying from 18 inches to 2 feet 3 inches, pierced with embrasures closely set, and almost rectangular in the opening, commanded almost every point outside; the crenelles, to which were attached hanging doors of wood moving upon a horizontal axis, and which were lowered or elevated by means of a notched iron quadrant bar, according as the enemy was more or less distant, allowed those within to reach easily the moats and surrounding country while themselves under shelter (fig. 47 [y]).

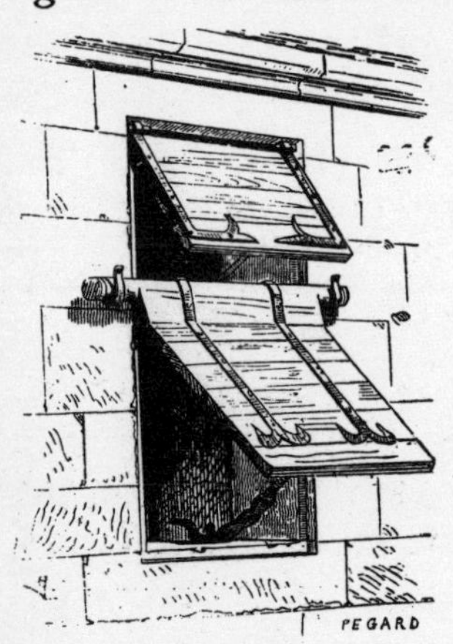

Fig. 47. A Crenelle with its wooden Hanging Shutter.

y We give a drawing of one of these crenelles of the upper stories of the towers of the city of Carcassonne, which date from the end of the thirteenth century. The

Round towers flanking the curtains resisted the action of the sap and the blows of the battering-ram better than square ones, and on this account had been adopted generally, from the first, in the fortifications of the Middle Ages; but at the close of the twelfth century they were of small diameter, and capable of containing a very restricted number of defenders; the limited extent of their circumference allowed of no more than two or three loopholes on each story, and they could therefore only operate feebly against the two adjoining curtains; their diameter was increased in the thirteenth century, when they were provided with stories down to the level of the moat. It was easier for the besiegers to batter a tower than a curtain (fig. 48), for when once established at the point

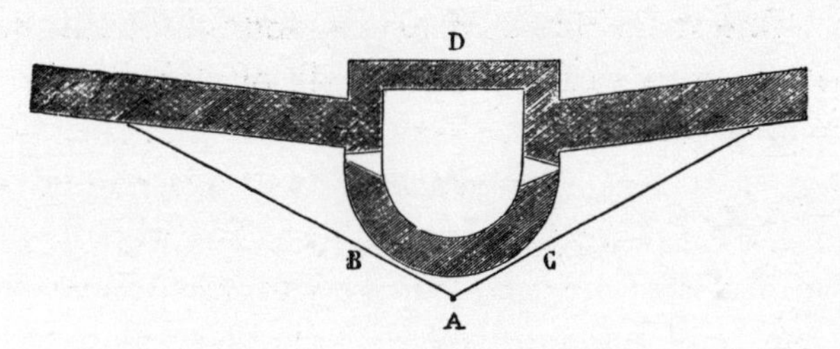

Fig. 48. Plan oɪ part oɪ a Curtain-wall with a Bastion.

A. Point of the Bastion. B C. Flanks of the Bastion, on which the Hoards were placed. D. Inner Wall of the Bastion or Tower.

A, and when they had succeeded in burning the hoards from B to C, the besieged had no longer the power to molest them; but in the enceintes of the towns, all the towers being closed at the gorge, D, when the assailants had made a breach at A, or thrown down the semi-

---

lower door, or louvre-board, hung simply from the two iron hooks in the wall, was removed when the hoarding was put up, as it was through these crenelles that the garrison passed from the interior to the hoards. As for the upper boarding, it was fixed permanently by means of hinges at either side of the opening, and could be raised to let in light and air without danger from the projectiles outside, when the lower portion was down.

circumference of the tower, they still had not effected an entrance into the town, but had new difficulties to overcome. This is the reason why they preferred, when laying siege to a fortified place, to attack the curtains, although their approaches were more difficult than those of the towers (fig. 49); for the besiegers when they had

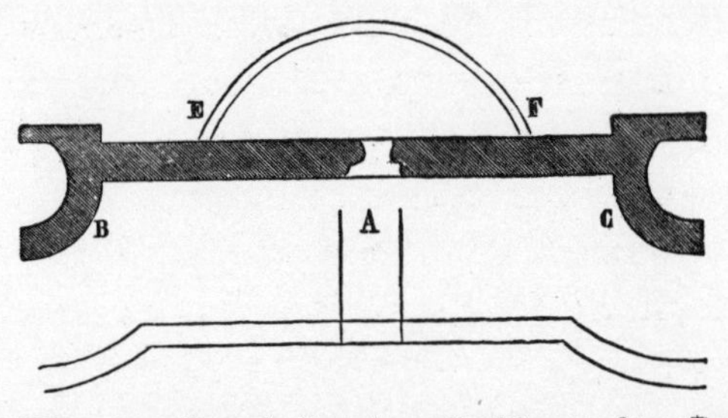

Fig. 49. Plan of one Bay of a Curtain-wall with part of two Bastions.

A. The Weakest Point, or Breach. B C. The Two Bastions.
E F. Temporary wall thrown up by the besieged within the Breach.

reached the point A, after having destroyed the upper defences of the towers, B C, and made their breach, were in the town, unless, as often happened, the besieged had rapidly thrown up a second wall, E F; but it seldom was found that these provisional defences could hold out for any length of time. The assailants, however, in all well-directed sieges, made simultaneous attacks, some by means of the sap, others by the mine, and others finally (these last being the most terrible) by means of moveable turrets; for, when once this turret had been brought close to the walls, the success of the attack was no longer doubtful. But in order to be enabled to bring these wooden towers, without risk of having them burnt by the besieged, close to the parapets, it was necessary to destroy the hoards and battlements of the adjoining towers and curtains, a labour which it required numerous engines and much time to effect. It was necessary to fill the moat completely, and to be certain, moreover, when the moat was filled, that the besieged had not mined its bed under the point upon which the tower was directed, an operation which they seldom left untried, if the nature of the soil did not present an insuperable obstacle.

So early as the close of the thirteenth century the necessity had been felt, in order more completely to command the curtains, not only of increasing the diameter of the towers, and thereby rendering the destruction of their upper defences a task of greater length and difficulty, but, further, of increasing their flanks by terminating them exteriorly with a projecting angle which already gave them the form of a horn (fig. 50). This

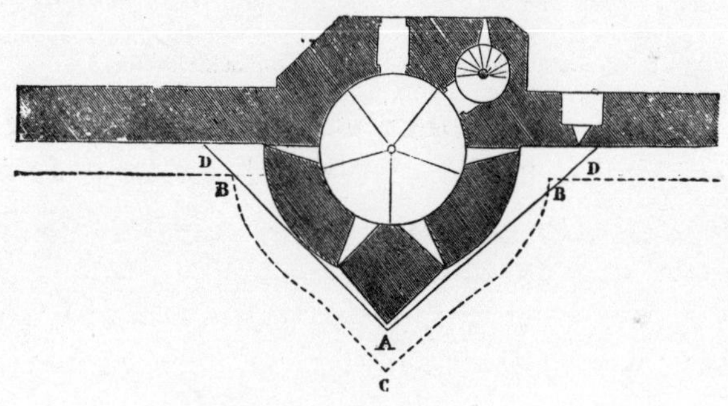

Fig. 50. Plan of a Horn.

A. The Beak. B B C. The Hoards of the Horn. D D. The Hoards of the Curtain.

angle had several advantages: firstly, it considerably increased the force of resistance of the masonry of the tower at the point where it would be likely to be attacked by the ram or the sap; secondly, it defended the curtains better by extending the flanks of the hoards, BC, which thus assumed the form of a line nearly perpendicular to the ramparts; and, thirdly, by keeping the pioneers at a distance, it allowed those placed in the hoards of the curtains at D to reach them at an angle much less acute than when the towers were circular, and consequently to hurl their projectiles from a less distance and with greater effect. At Carcassonne these projecting angles, or horns, are of the form shewn in plan by the figure 50. But at the castle of Loches, as well as at the gate of St. John at Provins, they were given the form, in plan, of two broken or intersecting curves

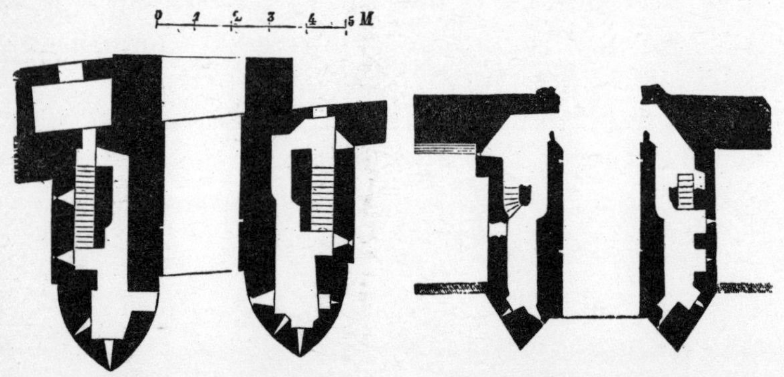

Fig. 51. Beaks of Loches and of the Gate of St. John at Provins.

Fig. 52. Beaks of the Gates of Jouy at Provins, and of Villeneuve-le-Roi.

(fig. 51), and at the gate of Jouy in the same town (fig. 52) and at the gates of Villeneuve-le-Roi that of rectangular works terminating in a point, in such a way as to command obliquely the entrance and the two adjacent curtains. It will therefore be seen that the inconvenience of circular towers had been discovered from the beginning of the thirteenth century, and the weakness inherent in them at the point of the tangent parallel with the curtains. The use of the means here indicated, however, appears to have been reserved for places very strongly defended, such as Carcassonne, Loches, &c.; for occasionally in places of the second order they were content to have square towers of slight projection for the defence of the curtains, as may be seen to this day on one of the fronts of the enceinte of Aigues-Mortes (fig. 53), the ramparts of which (with the exception of the tower of Constance, A, which had been built by

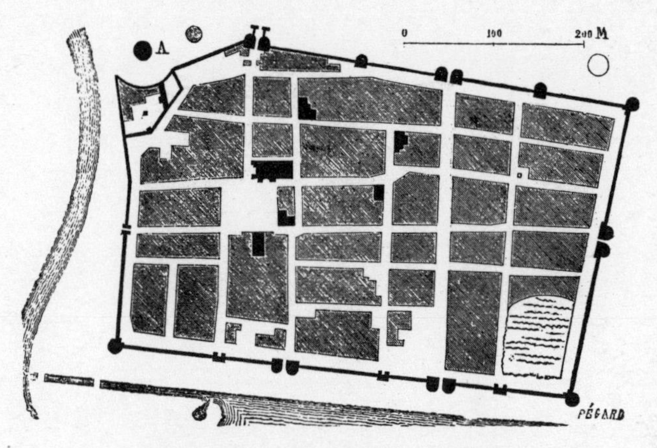

Fig. 53. Plan of the Town of Aigues-Mortes.

St. Louis, and which was used as a donjon and light-house) were erected by Philip the Bold[z].

But it was at the projecting angles of fortified places that the necessity more especially was felt of placing the strongest possible defences. As is still the case at the present day, the assailants looked upon a projecting angle as easier of access than a flanked front. The engines for hurling projectiles did not carry to any great distance until the use of cannon, and the salient

---

z "Philip the Bold, who quitted Paris in the month of February, 1272, at the head of a numerous army, to endeavour to take possession of the Comté of Toulouse, and to punish, on his passage, the revolt of Roger Bernard, Count of Foix, stopped at Marmande. There he signed, in the month of May, with William Boccanegra, who had joined him in this town, a treaty, whereby the latter engaged to furnish the sum of 5,000 livres tournois (about 3,500*l.* sterling) for the construction of the ramparts of Aigues-Mortes, in consideration of the cession made over to him and his descendants by the king, as a fief, of one half of the manorial taxes to which the town and port were subject. The letters-patent to this effect were countersigned, in order to render them more authentic, by the great officers of the crown. At the same time, and for the purpose of contributing to the same charges, Philip commanded there should be levied in addition to the denier in the pound already fixed, a fortieth part on all merchandize which should be brought into Aigues-Mortes by land or sea."—*Hist. génér. du Languedoc.*

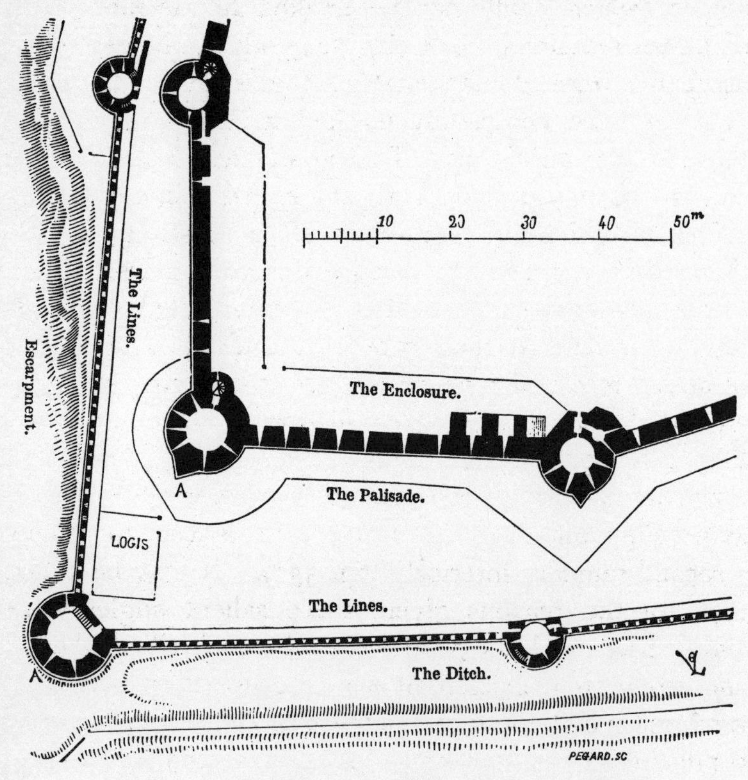

Fig. 54. Plan of an Angle of the Fortifications of Carcassonne.

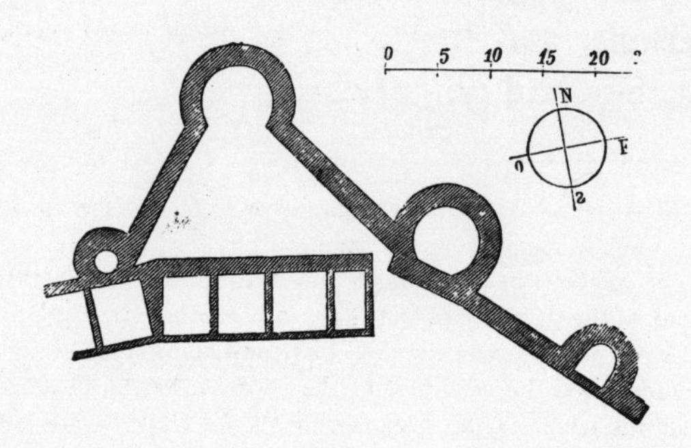

Fig. 55. Plan of a Projecting Angle of the Castle of Falaise.

angles (which could not be flanked by defences at a distance) remained weak (fig. 54); when, therefore, the assailants were able to establish themselves at the point A, they were completely masked as far as concerned the defences adjoining. Thus it was necessary that the corner-towers (*tours du coin*), as they were then generally called, should be in themselves of great strength. To this end they were built of a greater circumference than the others, and were raised to a greater height; the external obstacles at their base were multiplied by means of wider moats, by palisades, and sometimes even by advanced works; they were armed with projecting horns; they were isolated from the adjoining curtains; care was taken to make the two towers of the returns [a] as strong as possible, and sometimes these towers were united by a second rampart interiorly (fig. 55 [b]). It may be added that, for the reasons given, these salient angles were avoided as much as possible in all well-fortified places; and when they existed, it was because they had been rendered inevitable by the nature of the site, in order to command an enscarpment, a river, or a road, and to prevent the enemy from establishing himself on the dead level of the base of the ramparts.

Down to the fourteenth century the gateways were provided with gates strongly lined, with portcullises, machicolations, and brattishes of two and three stories high; but they had no drawbridges.

---

[a] The plan here given is that of the western angle of the double enceinte of the city of Carcassonne, built by Philip the Bold.

[b] This salient angle (fig. 57), which indicates clearly the arrangement above described, is one of the defences of the thirteenth century attached to the castle of Falaise. We have already seen how, at Château-Gaillard, Richard Cœur-de-Lion had perfectly understood the weakness of the corner-tower of his fortress, and how he had detached the whole of the projecting work from the castle proper by means of a double-flanked rampart and a moat.

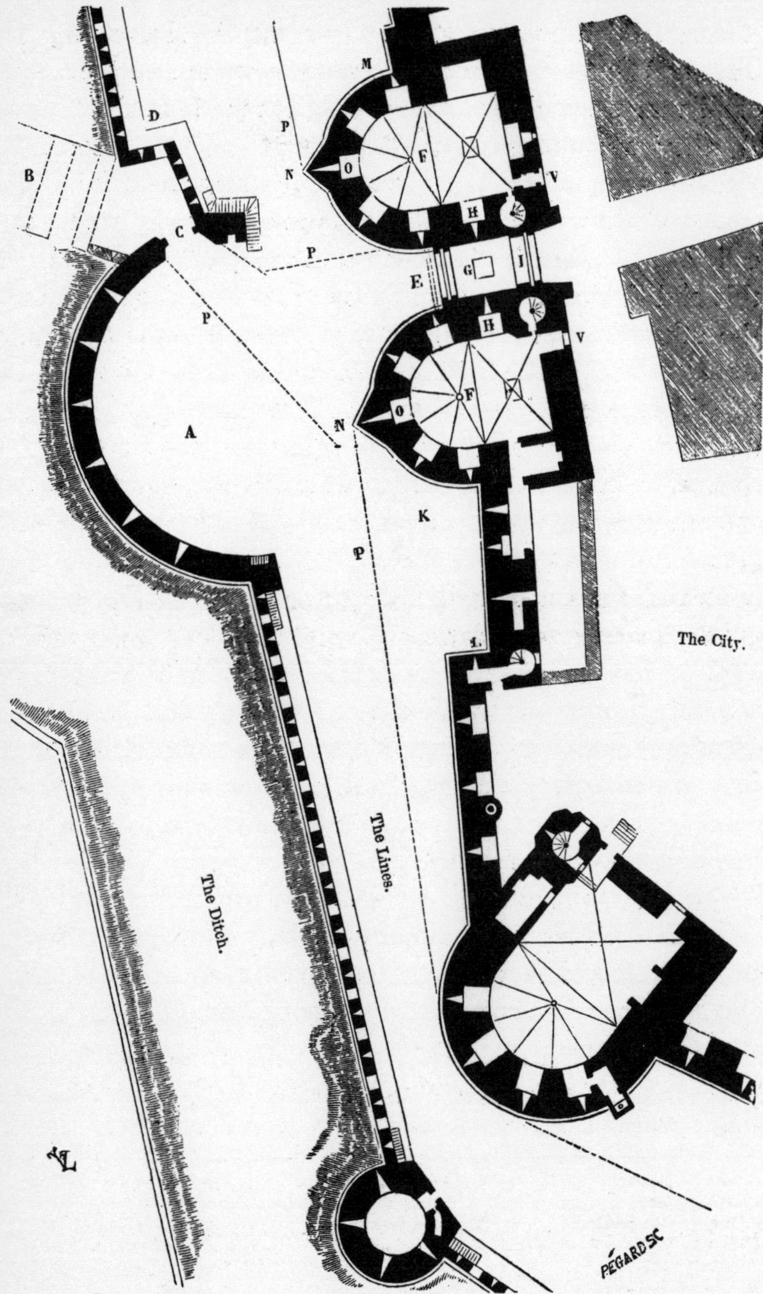

Fig. 56. Plan of the Narbonne Gate of the City of Carcassonne. (See opposite.)

The fine Narbonne gate of the city of Carcassonne (fig. 56), which is one of the strongest we are acquainted with, and the construction of which dates back to Philip the Bold, is not closed by a drawbridge. In front of the barbican, A, or the bridge, B, which spanned the moat, one or more moveable causeways orginally existed. This plan shews how the entrance, C, of the barbican was flanked by a *redent* (return) of the curtain D, and how care had been taken to mask it from those on the outside. If the assailants forced this first gate, they presented themselves in flank before the gate of the city, E. The passage between the two towers, FF, was closed, firstly, by a chain thrown from one tower to the other; secondly, by a machicolation; thirdly, by an outer portcullis; fourthly, by strong gates, solidly lined with iron, and furnished with heavy iron bars; fifthly, by a great square machicolation, G, and two loop-holes, H; sixthly, by a third machicolation placed in front of the second or inner portcullis, which was lowered at I. If the assailants presented themselves at K in order to sap the foot of the tower, they were taken in reverse by the redent L, surmounted by a large watch-tower pierced with loopholes; on the other side of the gate, at M, they were in the same way open to be attacked from a tower close at hand. As we have already explained, the projecting horns, N, forced the pioneers to unmask themselves before the neighbouring curtains; and loopholes, O, pierced at the ground level in the ground-floor chamber, opposed their approach. Palisades, P, had to be forced before they could sap the base of the walls, or place their

A. The Barbican.
B. The Bridge.
C. Entry to the Barbican.
D. Return (*Redent*) of the Curtain-wall.
E. Gate of the City.
F F. The Flanking Towers.
G. A Square Machicoulis (or Murdering-hole).
H H. Loopholes.
I. The Second Portcullis.
K. Part of the Curtain.
L. Another *Redent* of the Curtain.
M. Part of the Curtain.
N N. The Beaks of the Tower.
O O. Loopholes.
P P. Palisades.

ladders; and these palisades were, in the case of an attack, furnished with numerous defenders [c]. Over the entrance of the Narbonne gate, at E, were placed in times of war a brattish of wood pierced with loopholes, and with two heights of machicolations [d]. The plan of the first story of the Narbonne gate (fig. 57) is composed, firstly, of a central hall or chamber, in the floor

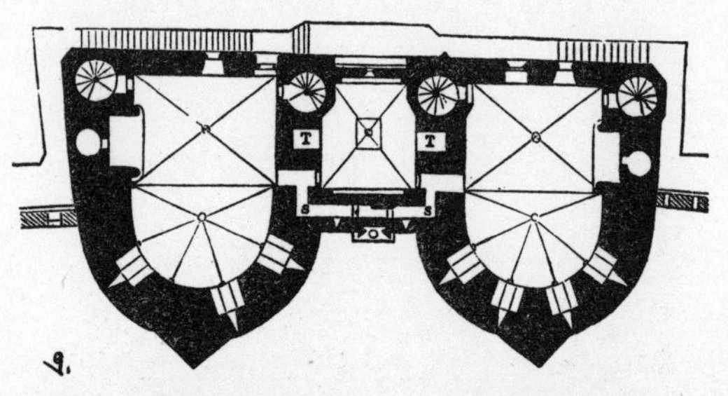

Fig. 57. Plan of the First Floor of the Narbonne Gate.

S S. Small Corridor over the Machicoulis. T T. Recesses in the Walls.

of which are pierced the great square machicolation, and the oblong machicolation in front of the second portcullis; it was also from this chamber that the first portcullis was drawn up or let down. Observe the two small corridors, S, with their elbows, or right-angled turnings, which serve as a communication above the first external machicolation, and which are so arranged as to allow the garrison to hurl their missiles upon the assailants without being themselves visible. On either side of the central machicolated chamber are two recesses, T, formed in the thickness of the wall, which served likewise to mask the defenders while engaged in rolling down materials on the assailants, when they had

[c] In the same plate we have given the plan of the first story of the tower called the *Trésau*, of which we already have had occasion to speak.

[d] The holes and corbels necessary to the laying of this brattish are still perfectly visible.

been arrested in their progress by the second portcullis. Secondly, of two halls or chambers in the two towers, furnished with fire-places containing ovens; this first

Fig. 58. Elevation of the Narbonne Gate.

story was vaulted like the ground-floor. From the ground-floor the way up to the first floor was by two staircases adjoining the passage; but in order to reach the second story and top battlements from the first story, it was necessary to take the two staircases at the external angles. It was in this manner that, at every step, obstacles and difficulties were multiplied; in cases of surprise it was necessary to be familiar with the localities in order not to lose one's way amidst so many turnings and means of access so carefully disguised. From the city, access to the tower could only be obtained through the two doorways, V, the parapets of the curtain-wall being at a great elevation above the level of the streets and without any direct communication with them. The two angle staircases led to the second story, which comprised the whole internal area of the work[e]. This story communicated on its anterior face, between the two towers, with the great brattish shewn on the elevation (fig. 58) through an opening, which, when the brattishes were removed in time of peace, served as a window[f]; it was amply lighted on the town side by five large pointed windows, with mullions, defended on the outside by strong cross-barred grilles. Finally we reach the battlemented top story which carried the roof (fig. 59). The great square machicolation (or trap), pierced at the centre of the passage in the vaulting of the ground-floor, was repeated in the vault over the first story, and also, probably, in the floor of the embattled story. Through this trap, which served as a means of defence, orders could also be communicated from the upper portions of the gate to the lower stories; for, according to the arrangement of the towers of this period, all orders

[e] In the fifteenth century partitions were constructed to separate this hall into three.

[f] In this geometric elevation we have drawn the brattish and hoards as in position at the top of one tower.

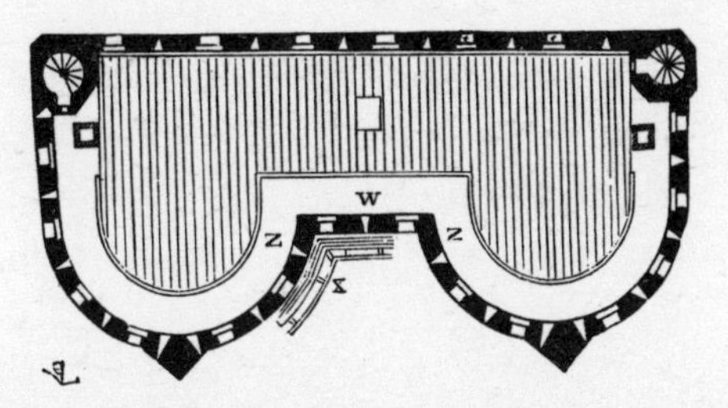

Fig. 59. Plan of the Upper Story of the Narbonne Gate.

X. Part of the Hoarding shewn in the Elevation, fig. 58. W. Hoarding in Front.
Z Z. Hoarding on the Flanks.

would be given from the upper stories, inasmuch as therein was concentrated the active element of the defence. The whole of the anterior portion of this gateway could be furnished at the level of the crenellated story with hoards, the holes for which are all existing, pierced at regular intervals, at the base of the parapets. In the plan (fig. 59) and in the elevation (fig. 58), we have shewn at X a portion of these hoards in position. The brattishes and hoards could easily contain two hundred men, without reckoning those whose duty it was to bring up and distribute the projectiles, and who did their work without interfering with those in the outer works. If the brattish were taken by escalade, or destroyed by the missiles of the enemy, the single opening which gave access to the interior was closed, and the assailants were exposed to the fire of the two flank hoards, Z, and of the front, W. The second portcullis was manœuvred from the outside parapet, and orders were communicated to those actively engaged within the work by means of a small barred window, which looked into the central chamber of the first story, about the height of a man from the floor. From the ground-floor of the two towers you descend to two cellars (covered with quarter-spherical and barrel-vaulting) by means of two traps, closed over by flags. The whole appearance of this work corresponds with its effective force; the walls are built, in large courses, of a greyish sandstone of great hardness; all the external wall-faces are rusticated, that is to say, the joints of each stone are relieved by a sinking, and the middle of the stone is left rough. This kind of masonry was much in use at the close of the thirteenth and beginning of the fourteenth century, for fortifications. It is in this way that all the curtains and towers of the city of Carcassonne and the ramparts of Aigues-Mortes, dating from the reign of Philip the Bold, are executed.

In castles it frequently happened that wooden drawbridges, which were removed in time of siege, completely

Fig. 60. The Drawbridge.

intercepted the communication with the outside; but in the enceintes of cities, the approaches were defended by palisaded barricades, or by barbicans; and these barriers having been once carried, troops could ordinarily enter the city on the level. It was not until the commencement of the fourteenth century that the practice began to be adopted of fixing at the entrance to the bridges thrown across the moats before the gates of the town, drawbridges of wood in connection with the barriers (fig. 60), or with advanced works in masonry (fig. 61 [g]);

Fig. 61. Entrance to the Castle of Montargis.

and, finally, in a little while, that is to say towards the middle of the fourteenth century, the drawbridge was applied to the gates themselves, as may be seen at the fort of Vincennes, amongst other examples. We must, however, add that in many cases, even during the fourteenth and fifteenth centuries, the drawbridges were merely attached to the advanced works. These draw-

---

[g] **Entrance to the castle of Montargis, on the side of the road from Paris to Orleans, (Ducerceau *Châteaux royaux en France*).**

bridges were constructed in the same manner as those generally adopted at the present day; that is to say, the bridge was composed of a causeway or platform of wood,

Fig. 62. The Tapecu, or Shutter suspended from above.

which moved upon an axis, and was raised and lowered by means of two chains, levers and counter-weights;

Fig. 63. A Shutter balanced on a Pivot.

when raised, the causeway closed (as it still closes in our fortresses) the archway of the passage. But there were other kinds of moveable doors employed during the twelfth, thirteenth, and fourteenth centuries; there was the *tapecu*, especially intended for posterns, which, revolving on an axis placed horizontally at the top of the hanging door, fell back as the person went out (fig. 62); and the gates of barriers which revolved on axes placed about the middle of their height (fig. 63), one of the two halves serving as a counter-weight to the other. In the

Fig. 64. Gate of Aubenton, attacked by the Count of Hainault, from a MS. of Froissart.

fine manuscript of the Chronicles of Froissart in the Imperial Library (of France[h]), we find a vignette which

[g] Manusc. 8320, vol. i. in-fol., beginning of the fifteenth century. This vignette, of which we here give a portion, accompanies chap. xlvi. of this manuscript, en-

represents the attack on the barriers of the town of Aubenton, by the Count of Hainault. The gate of the barrier is defended in this manner (fig. 64); it is provided with two wooden towers of defence. Behind, we see the gate of the town, which is a stone building, although the text describes the town of Aubenton as "only closed with palings." Soldiers are in the act of flinging from the battlements a bench, and other pieces of furniture and of pottery.

We have seen how during the twelfth and thirteenth centuries it was customary to protect the summits of towers and curtains by timber hoarding. It is unnecessary to say that the assailants endeavoured, by means of their catapults and other machines of that nature, to shatter those hoards with stones, and to burn them with inflammatory projectiles; a result which they easily obtained, if the walls were not of a great height, or if the hoards were not covered with raw hides, or were not *hourdés*, that is, coated with loam or mortar. Already, towards the middle of the thirteenth century, an attempt had been made to render the timber hoards less liable to be burnt by resting them on corbels of stone. It was in this way that, at Coucy, the hoards of the town gates,

---

titled, *Comment le comte de Haynault print et détruit Aubenton en terrasse.* It forms chap. cii. of the edition of the *Chronicles* in the *Panthéon littéraire*. . . . . . "Then began the assault with terrible force, and the crossbows were set to work, inside and out, to keep up a vigorous fire; by which fire were many wounded, both of defendants and assailants. The Count of Hainault and his guard (*route*), wherein were many good knights and squires, rode up to the barriers of one of the gates. Thereat was there an assault fierce and terrible. On the bridge likewise, at the gate towards Chimay, where were Messire Jean de Bourmont and Messire Jean de la Bove, there was another terrible assault and fierce conflict, and the French were obliged to withdraw within the gate; for they lost their barriers and the bridge also, which were taken by the Hainault men. And the assault was dire, for those who were mounted on the gate hurled down timber and planks, and earthen pots filled with lime, and a great quantity of stones, wherewith they wounded and crushed all those who were not covered with strong armour."

and of the towers and donjon, which belong to this period, were supported (see fig. 45). Still the faces and floors of these hoards might take fire. In the fourteenth century, during the wars of this period, when so many towns were given up to fire and pillage, "arses et robées," as Froissart has it, the timber hoardings were almost everywhere replaced by continuous brattishes of stone, which possessed all the advantages of the hoards, inasmuch as they commanded the foot of the walls, without any of their inconveniences: these new crest-works could not be burnt, and offered a greater resistance to the projectiles hurled by the engines; they were stationary, and not fixed only in time of war, like the wooden hoards. We have seen in the case of Château-Gaillard how Richard Cœur-de-Lion had already applied, in a manner in advance of his century, an excellent and incombustible kind of machicolation. But in order to secure a broad parapet

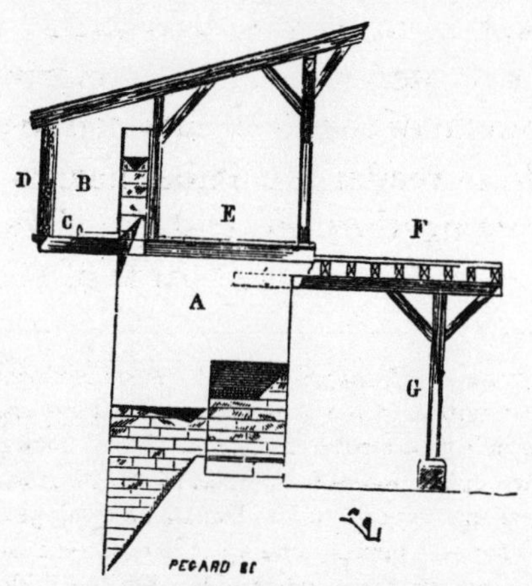

Fig. 65. Plan of the Hoarding.

A. The Allure, or Walk on the Stone Wall. B. An outer Gallery of Wood. C. Machicoulis. D. Loopholes. E. Upper Gallery. F. Inner Gallery. G. Post to carry it.

for the defenders, and a projection from the face of the walls which allowed of opening machicolations of a good size, it soon became necessary to modify the whole system of construction in the upper portions of the defence. By means of wooden hoards there was added to the parapet-walk in masonry, A (fig. 65), a projecting gallery, B, having pierced machicolations at C, and loopholes at D; but the width of the parapet was often further increased, either by extending the hoards interiorly in the direction of the town at E, or by adding to the parapet-walk wooden joisting, F, the beams to support which were let into cavities formed, at regular intervals, under the top of the curtain-walls; which beams were supported at their other extremity on story-posts, G. These supplemental defences were generally reserved for such curtains as appeared weak and of easy approach[i]. Hoards had the advantage of allowing the stone parapet to remain, and of preserving intact behind them another system of defences, when they were burnt or otherwise injured. It was with difficulty that, with stone brattishes and machicolations, those wide spaces and those divisions so useful to the defence could be obtained; we will describe the measures which were taken for curtain-walls, which it was considered important to have strongly defended (fig. 66). Corbels were laid in courses, one projecting over the other, at intervals of from 2¼ to 4 feet at most, from centre to centre. On the outer extremity of these corbels was erected a pierced parapet, B, 12 to 16 inches thick, of stone, and about 6½ feet high. In order to counterpoise the overhanging corbels at C, an inner wall was erected, pierced with doorways and square apertures at regular distances, and of a sufficient height to afford the roof-covering the proper inclination. Behind

[i] At Carcassonne, on the south side, the ramparts of the second enceinte were furnished with these works in times of war; traces of them are perfectly preserved in the Narbonnaise gate, at the western angle-tower.

Fig. 66. Sections of part of a Curtain-wall, well defended.

B. The pierced Parapet, or Battlement. C. The Wall with Corbels. D. The Roof.
L. Wooden Gallery within the Wall. G. The Allure and Station for Archers.

the wall, C, were fixed wooden galleries, L, which took the place of the galleries, E, of the wooden hoards (fig. 65), and which were necessary for keeping the parapets supplied with projectiles and for the free passage of those engaged in that duty, without interfering with the archers and others posted at G (fig. 66). For towers, the arrangement was still more complete. Retaining the same disposition in the machicolated story as described for curtain-walls, they super-imposed on the wall, C, another story, H, pierced with crenelles or loops; and occasionally even at the base of the roof at I, another uncovered line of battlements was formed. So

Fig. 67. Battlements and Machicoulis of a Tower.

B. The Parapet. C. The Wall. G. The Allure on the Machicoulis. H. Upper Story. I. Upper Battlement. K. The Doors.

that were the covered way, G, to be taken by escalade, or by means of the moving turrets, after the destruction of the parapets, B; by barricading the doors, K, those holding the tower would still be able to drive back the assailants (who would thus find themselves hemmed in at G on a space without any issues) by flinging on them from the stories H and I, stones, beams, and all kinds of projectiles. The manuscript of Froissart in the Imperial Library (of Paris), which we have already quoted, gives a great number of towers arranged in this manner amongst its vignettes [k]. Many of these drawings shew

Fig. 68. Newcastle-on-Tyne, from a MS. of Froissart.

A. The Hoarding.

that the timber-hoards, A, were retained, together with the stone machicolations, the former being kept for the defence of the curtain-walls; and, in point of fact, those two modes of defence were long applied together, the brattishes and hoards of wood being much less costly in the erection than stone machicolations. The castle of Pierrefonds, built during the latter years of the four-

[k] Vignette accompanying chapter cxxv., entitled:—"How King David Bruce of Scotland came with his whole army before the new castle on the Tyne."

teenth century, still displays, in a very complete manner, those two kinds of upper defences. We give (fig. 69)

Fig. 69. Part of the Castle of Pierrefonds.

A. The Machicoulis.
B. Tail Stones of the Parapet.
C. Weather-moulding of the Roof.
D. The Allure.
E. Corbels of the Roof.
F. Openings in the Wall.
G G. The Doors.
H. Upper Story.
I. Upper Battlement.
K. The Stair-turret and Watch-tower.

the present state of the angle formed by the north-

Fig. 70. Part of the Castle of Pierrefonds, ***restored.***

western tower and the north curtain-wall. We see perfectly, at A, the machicolations still in position; at B, the tailing of the stone parapets where they entered the wall of the tower; at C, the weather-moulding of the shed-roof which covered the parapet-walk, D; at E, the stone corbels which carried the ridge of this roof; at G, the doors which communicated with the parapet from the staircase, and at F, openings which served for passing projectiles from the interior of the tower to those defending the battlements; at H, an embattled story, opening above the machicolations; at I, the last uncovered battlement at the base of the roof: finally, at K, the staircase-tower, used as a watch-tower at its summit. But, in castles more particularly, because of the smallness of the reserved space between their enceintes, the curtains served as external walls to the buildings placed between the towers along the line of those enceintes, so that the parapet-walk gave access to the chambers which thus occupied the place of the wooden shed, L, shewn in figure 66. We subjoin (fig. 70) the restoration of this portion of the defences of Pierrefonds. From this the destination of each of the details of military construction which we have just described will be easily understood. But in this case we have the strongest possible forms of defence which were adopted for walls and towers; many works were inferior to those as to arrangement, and were composed merely of battlements and machicolations of slight projection, with narrow parapet-walk. Such are the walls of Avignon, which, considered as to their preservation, are certainly the finest at present existing in French territory; but which, looking at them with a view to their effective strength, did not present a formidable defence for the period at which they were erected. Following the method then in use in Italy, the walls of Avignon are flanked by towers which, with some exceptions, are square [1]. In France the round tower had been considered, and justly, as stronger than the square one; for, as we have already demonstrated, the pioneer, while engaged at the base of the round tower, was commanded obliquely by the adjoining curtains, whilst if he attained at the base of the external face of a square tower, at O, he was completely covered as regarded the defences in his immediate proximity (fig. 71); and by preventing those defending from shewing themselves at the battlements, and by the destruction of a few of the machicolations immediately over him, he might pursue his sap-works in perfect security. Contrary also to the rules of French fortification in the thirteenth and fourteenth centuries, the square towers of the ramparts of Avignon are open on the city side (fig. 72); and, consequently, no longer tenable from the moment the enemy had obtained an entrance into the city. The walls of Avignon are more than a flanked enceinte, representing the external en-

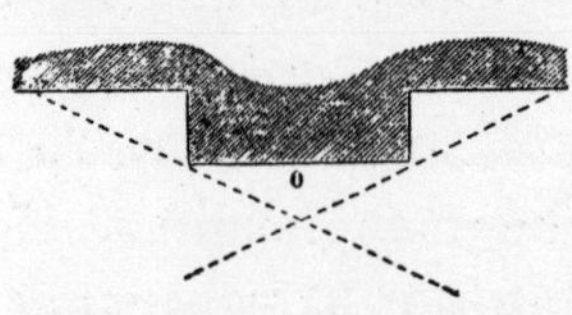

Fig. 71. Plan of a Square Tower.

[1] We have already seen that the ramparts at Aigues-Mortes are likewise, upon one front, flanked by square towers, and we should bear in mind that they were erected by the Genoese, Boccanegra. The enceinte of Paris, however, which was rebuilt under Charles V., was likewise flanked by oblong towers; but the enceinte of Paris never was considered as of any great strength. Square towers belong rather to the south than the north of France; the ramparts of Cahors, which date from the twelfth, thirteenth, and fourteenth centuries shew square towers, of a fine arrangement for defence; the ramparts of the towns of the *comtat* of Venaissin are mostly furnished with square towers which date from the fourteenth century, as well as the greater number of the towns of Provence and the Rhône. Orange was provided with square towers, constructed at the close of the fourteenth century. The Normans and Poitevins, up to the time when the provinces were united to the *Domaine Royal*, that is to say, until the beginning of the thirteenth century, appear to have adopted, in preference, the square form for their towers and donjons. The majority of the ancient castles built by the Normans in England and Sicily contain rectangular defences.

Fig. 72. Part of the Walls of Avignon, inside.

ceintes of towns having a double line of fortification, and not curtain-walls broken at intervals by forts which could themselves hold out against an enemy when master of the place. These walls are not even furnished with machicolations throughout their whole extent, and the south side of the city is only defended by simple battlements, which were not intended to receive wooden hoards. Their height does not everywhere reach the minimum given to good defences in order to place them beyond the reach of the scaling ladders (*échelades*[m]).

---

[m] Escalade by means of ladders.

There is, nevertheless, in the enceinte of Avignon, a certain grandeur of arrangement, an unity of combination, which proves that at this period the art of fortification was complete, that method had taken the place of hap-hazard experiment in the defence of cities, and that the constructors were guided by the light of experience and a long practice. In order to complete our description of the defences of Avignon we give here examples of the system of flanking generally adopted. Fig. 73 shews the ground-story of one of these towers;

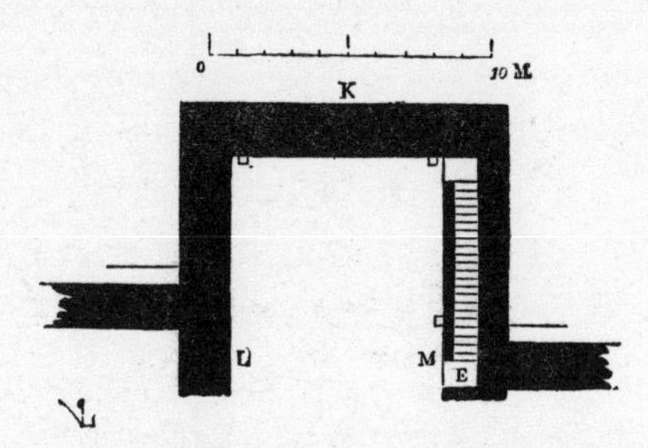

Fig. 73. Ground-plan of one of the Towers of Avignon.

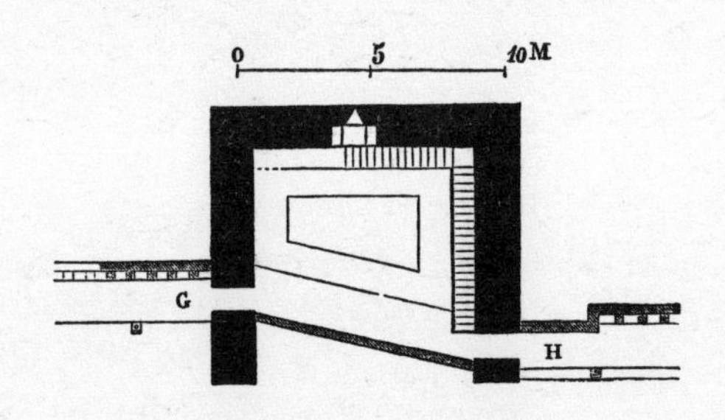

Fig. 74. Plan of the First Story.

E. Staircase leading to G H. Curtain adjoining. K. The plan for a Breach.
L M. The plan for an Inner Wall.

a staircase, E, closed by a door, gives access to the first story, which communicates by means of two doors with the adjacent curtains, G H. A second staircase projected on corbels leads up to the battlements (fig. 75), which are pierced with machicolations. This tower is

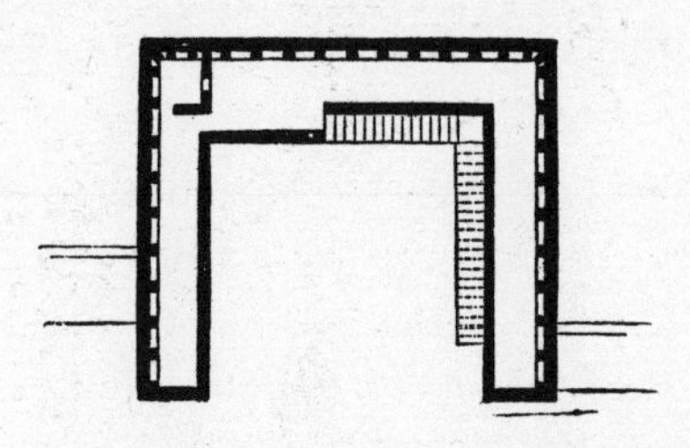

Fig. 75. Plan of Upper Story with the Allure and Battlement.

incapable of defence, it will be observed, except at its summit. The perspective view (fig. 76), taken from the city side, explains perfectly the whole system of defence, and the means of access to the several stories. This tower is one of the strongest of the place and is not closed towards the town; it allowed a considerable number of men to muster on the top parapets, and if it should be sapped at the point K (fig. 73) by the besiegers, it was still possible to defend the breach either by throwing a rampart across from one flank wall to the other, from L to M, or by hurling stones on the assailants through the great machicolation, pierced in the floor of the first story. But if the city walls of Avignon present only a defence of the second or third order, the castle, which was the residence of the popes during the fourteenth century, was a formidable citadel, capable from its site, its extent, and the height of its towers, of sustaining a long siege. There, again, the towers are square, but of such a height and thickness as to be able to defy the sap, and all projectiles hurled from the engines then in use; they are crowned by parapets and machicolations resting on corbels. The machicolations of the curtains are composed of a series of pointed arches,

Fig. 76. Perspective View of the Interior of one of the Towers of Avignon.

leaving behind them and the external face of the wall a space adapted for hurling stones or any other kind of projectile through. In the provinces of the south and west this kind of machicolation was much used, and they were certainly preferable to the machicolations of timber hoards, or stone parapets resting on corbels, for the reason that they were continuous and not interrupted by beams or stone consoles, and allowed consequently long and heavy pieces of timber to be hurled down on the assailants, along the face of the wall (fig. 77); which,

Fig. 77. Part of the Palace of the Pope at Avignon.

falling obliquely, were sure to crush the cats or shields (*pavois*), under which the pioneers were lodged.

The art of fortification, which had made great progress at the commencement of the thirteenth century, but which remained almost stationary during the course of that century, again began to progress during the wars which took place from 1330 to 1400. When Charles V. had restored order to his kingdom, and had retaken a large number of places from the English, he caused nearly all the defences of the reconquered towns and castles to be either repaired or rebuilt; and, in these new defences, it is easy to see a method and a regularity which indicate an art already in a state of advancement, and based upon fixed rules. The castle of Vincennes is an example of what we here advance (fig. 78 [n]). Built in the plain, there were here no particular circumstances of site to be attended to; and we find accordingly that its enceinte is perfectly regular, as likewise the donjon and its defences. All the towers are oblong or square, but lofty, massive, and well-furnished at their summits with projecting bartizans flanking the four faces; the donjon is likewise flanked at the angles by four turrets; the distances between the towers are equal; these latter are closed and capable of separate defence [o]. The castle of Vincennes was begun by Philippe de Valois, and finished by Charles V., with the exception of the chapel, which was only terminated under Francis I. and Henry II.

The feudal system was essentially adapted for the

---

[n] We subjoin the plan of the castle of Vincennes, which is rather a large *place d'armes*, or fortified enceinte, than a *castle* in the ancient meaning of the word. At E E are the two only entrances within the enceinte; these were defended by advanced works and by two lofty oblong towers: at A is the donjon, enclosed by an enceinte of its own, and a *chemise*, B. A moat, revetted and of great width, C, protects this donjon. At K are the ditches of the enceinte, the counterscarp of which is revetted, and has always been so. F is the chapel; G, the treasury; D, the bridge which gives access to the donjon; H and I are quarters for the garrison and stables. (See *Vue des maisons royales et villes*, Israël Sylvestre, in-f°.) We have taken from the plan given by Sylvestre only the buildings anterior to the sixteenth century. During the fourteenth and fifteenth century many others must have existed, but we know neither their situation nor their form.

[o] The smaller side of the parallelogram of the enceinte, including the projection of the towers, is 212 *mètres* long—695½ feet.

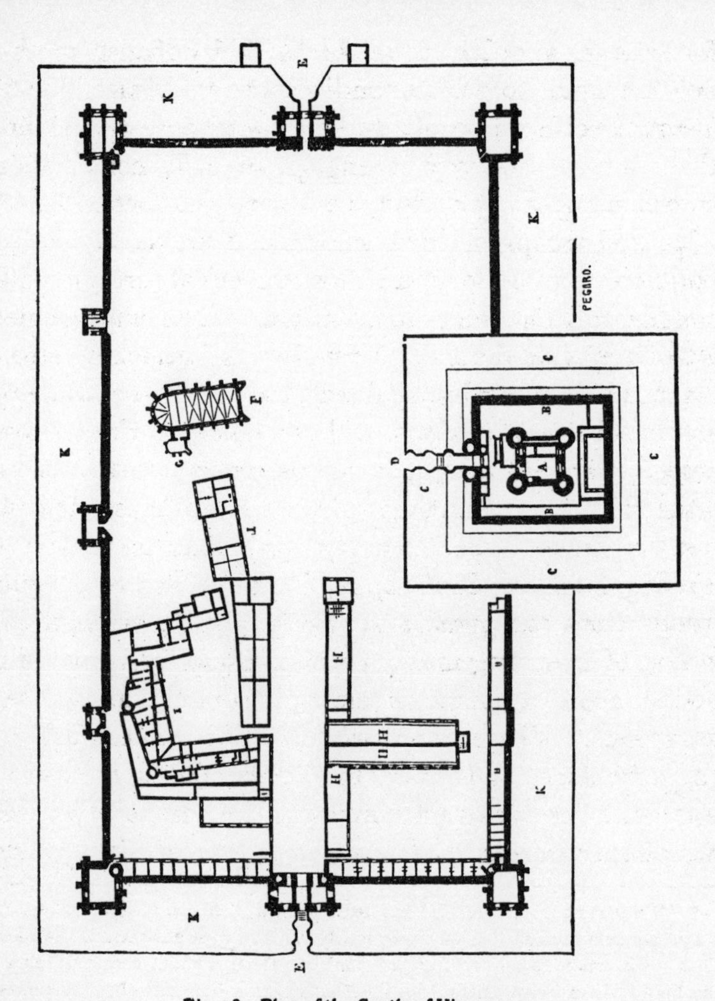

Fig. 78. Plan of the Castle of Vincennes.

A. The Keep or Donjon.
B. The *Chemise.*
C C. Ditch of the Keep.
D. Bridge to the Keep.
E E. Gates of the Castle.
F. The Chapel.
G. The Treasury, or Sacristy.
H H & I I Barracks and Stables.
K K. The Moat, or outer Ditch of the Castle.

defence and attack of strong places:—for the defence, for the reason that the nobles and their followers lived continually in these fortresses, which protected their life and their possessions, and were constantly intent upon improving them, and rendering them every day more formidable, in order to be able to bid defiance to their neighbours or to dictate conditions to their suzerain; for attack, because, in order to seize upon a castle in those times it was necessary to engage in daily conflicts, and consequently be able always to bring into action a body of picked troops of tried valour, and whose vigour and boldness counted for more than numerical force or skilful combinations in the plan of attack. The improvement introduced into the art of defending and attacking strong places was already highly developed in France, while the art of field-warfare was still stationary. France possessed excellent troops, men brought up to the use of arms from their childhood, brave to rashness, but she had no armies; her infantry was made up merely of hireling Genoese, Brabançons, Germans, and of irregular troops from the *good cities*, badly armed, without any notion of executing manœuvres, undisciplined, and, in an action, more a source of embarrassment than any real assistance. These troops were thrown into confusion at the first shock, and then they precipitated themselves upon the reserves and threw the squadrons of men-at-arms into disorder[p]. The passage from Froissart which

[p] "There is no man, unless he had been present, that can imagine, or describe truly, the confusion of that day; especially the bad management and disorder of the French, whose troops were out of number. What I know, and shall relate in this book, I have learnt chiefly from the English, who had well observed the confusion they were in, and from those attached to Sir John Hainault, who was always near the person of the king of France. The English, who were drawn up in three divisions, and seated on the ground, on seeing their enemies advance, rose undauntedly up, and fell into their ranks. That of the prince was the first to do so, whose archers were formed in the manner of a portcullis or harrow, and the men-at-arms in the rear. The earls of Northampton and Arundel, who commanded the second division, had posted themselves in good order on his wing, to assist and succour the prince, if necessary. You must know that these kings, earls, barons, and lords of France, did not advance in any regular order, but one after the other, or any way most pleasing to themselves. As soon as the king of France came in sight of the English, his blood began to boil, and he cried out to his marshals, 'Order the Genoese forward and begin the battle, in the name of God and Saint

we give in a foot-note *in extenso*, shews clearly what, during the first half of the fourteenth century, a French army was, and how little the noblesse thought of these troops of *bidauds*, or *brigands*[q], of Genoese bowmen,—in fact, of the infantry. The English began at this period to bring into the field an infantry which was numerous, disciplined, skilled in the use of the bow[r], and even already supplied with fire-arms[s]. The superiority of the *chevalerie* (or cavalry), which up to this time had been incontestable, was in its decline; the French *gendarmerie* went on sustaining defeat after defeat until the moment

---

Denis.' There were about fifteen thousand Genoese crossbow-men; but they were quite fatigued, having marched on foot that day six leagues, completely armed and with their crossbows. They told the constable they were not in a fit condition to do any great things that day in battle. The earl of Alençon coming to hear these words, was enraged and cried out, 'This is what one gets by employing such scoundrels, who fail you at your need.' . . . . When the Genoese were somewhat in order, and approached the English, they set up a loud shout, in order to frighten them; but they remained quite still and did not seem to attend to it. They then set up a second shout and advanced a little forward; but the English never moved. Yet a third time they shouted, loud and clear, then advanced within shot, strung their crossbows and began to shoot. Then those English archers advanced a step forward, and let fly their arrows in so dense a shower upon the Genoese that it was like snow. The Genoese, who were not accustomed to meet with such archers as those of England, when they felt the arrows piercing through heads, arms, and breasts, and through their armour, were sore discomfited; some of them cut the strings of their bows, and some flung them on the ground; and so they fell back.

"The French had a large body of men-at-arms on horseback and richly accoutred, to support the Genoese, who, when they wanted to retire, could not, for the king of France seeing them thus fall back, discomfited, rashly exclaimed, 'Kill me those scoundrels, for they stop up our road without any reason.' You should then have seen these men-at-arms lay about them on these runaways, of whom many fell never to rise more. And the English still kept shooting wherever there was the thickest press, and none of their shots were thrown away, for they struck the bodies of men and horses, who thereupon staggered and fell, and none could be raised up again without great trouble and the efforts of many men. And thus began the battle fought between Broye and Crécy in Ponthieu, on the same Saturday at the hour of vespers."

[q] So called because they wore a coat of mail called a *brigantine*.

[r] See *Etudes sur le passé et l'avenir de l'artillerie*, by Prince Louis-Napoleon, vol. i. p. 16 and following pages.

[s] At Crécy.

when Du Guesclin organized companies of tried and disciplined foot-soldiers, and by the ascendancy of his merits as a captain, succeeded in giving a better direction to the valour of his horse. These transformations in the composition of armies, and the use of cannon, necessarily modified the art of fortification,—slowly, it is true, for feudalism accommodated itself with difficulty to any innovation in the art of war; it was necessary for a long and cruel experience to teach it, to its cost, that valour alone was not sufficient for winning battles or taking towns; that the strong and lofty towers of its castles were not impregnable to an enemy who proceeded with method, spared his men, and took time enough in making his approaches. The war of sieges during the reign of Philip de Valois is not less interesting to study than the war of campaigns: in the one warfare, as in the other, the organization and discipline of the English troops gave them an incontestable superiority over the troops of France. Within the space of a few months the French army, under the command of the Duke of Normandy[t], lays siege to the fortified place of Aiguillon, situate at the confluence of the Lot and the Garonne, and the King of England besieges Calais. The French army, which was numerous (Froissart computes its strength at nearly one hundred thousand men), composed of the flower of our chivalry, after numerous assaults and feats of valour unparalleled, can make no impression upon the fortress; the Duke of Normandy, having lost many of his men, decides on undertaking a regular siege:—

"On the day after" (the unsuccessful attack upon the castle) "there came to the Duke of Normandy two master engineers,

[t] Son of Philip de Valois, taken at Poitiers: afterwards King John.

and said—'Sir, if you will let us have plenty of timber and workmen, we will make four great *Kas* [u], strong and high, upon four great and strong ships, which *Kas* or towers shall be brought close to the walls, and shall be high enough to overtop them.' The duke listened willingly to this offer, and commanded that those four towers should be made, whatsoever they might cost, and that there should be set to work all the carpenters of the country, who should be paid a good day's wages, to make them work the harder. These four towers were made after the plan and directions of the two masters, on four large ships; but they were long a-building and cost great sums of money. When they were complete, and the men had been placed in them who were to attack those in the castle as they had crossed the halt of the river, those last-named fired off four *martinets* [x], which they had recently had made to oppose the four towers already described. These four martinets flung huge stones, and the towers were so often struck by them, that they were soon shattered and broken, so that the men-at-arms, and those who impelled them, could find no shelter. So they were obliged to retreat; and in doing so, one of the towers foundered and was sunk in the river, and the greater number of those within it were drowned; which was a sad and pitiful thing, there being within it many good knights and squires who were eager to win honour for their names [y]."

The Duke of Normandy had sworn to take Aiguillon, nor durst any one in his camp even speak of raising the siege; but the counts of Guines and of Tancarville went to the king at Paris:—

"They related to him the present state and condition of the siege of Aiguillon, and how the duke his son had assailed it on many occasions, but had not been able to take it. The king was thereat struck with wonderment, but did not recall the duke his son; but desired rather that he should remain before

---

[u] The sequel shews those to have been towers, or *chaz-chateilz*.

[x] *Martinet*, an engine working by counterpoise, adapted for hurling great stones.

[y] Froissart, chap. ccxxvi. edit. Buchon.

Aiguillon, until he had succeeded in capturing and conquering it by famine, since by assault he could not take it."

No such rash imprudence marks the conduct of the King of England; he disembarks at La Hogue, at the head of an army, not numerous, but well disciplined; he marches through Normandy, taking care to have the main body of his army flanked by two bodies of light troops, commanded by captains acquainted with the ground who scoured the country right and left, and who every evening pitched their tents round about him. His fleet followed the line of march along the coast, so as to secure his retreat in the event of a check; and after each town taken, he sent the booty which it yielded on board his ships. He arrives, finally, at the gates of Paris; continues his victorious course as far as Picardy, where he is at last met by the army of the King of France, which he defeats at Crécy, and presents himself before Calais:—

"When the King of England came first before the town of Calais, in the manner of one who was determined on taking it, he besieged it upon a great scale and plan, and he commanded to be built between the town and the river, and the bridge of Nieulay, *hôtels* and houses, which were constructed of timber frame-work and orderly set in rows and streets, and the said houses were covered with thatch and broom, as though he intended to remain there for ten or twelve years; for it was truly his intention not to stir from before the town, winter or summer, until he had taken it, whatever time or pains it might cost him. And there were, in this new town built for the king, everything required for an army, and more besides; and a place was set apart for holding a market every Wednesday and Saturday, and there were mercers' ware, butchers' meat, cloth stores and all other necessary things: and each man might have what he willed for his money: and the whole of these matters came to them every day, by sea, from England and likewise from

Flanders, which countries supplied them with provisions and merchandize. With all this, the King of England's men scoured the country round about, the comté of Guines, and Therouenois, and as far as the gates of Saint-Omer and Boulogne, bringing back to the army great store of provisions of all kinds. *And the king did not make his people deliver any assault upon the said town of Calais, for he well knew that he would spend his pains and labour in vain.* Therefore he spared his men-at-arms and his artillery, and said that he would starve them out, however long a time it might take him, if King Philip of France did not appear a second time to encounter him and raise the siege."

King Philip arrives before Calais with a fine army; and the King of England at once has the only two passages by which the French could attack him, guarded. One of these passages was by the sand-hills along the sea-shore; the King of England has—

"all his ships and boats drawn up opposite these sand-hills, and well furnished with bombards, crossbows, springalds, and all such things: so that the French host neither dared nor were able to pass."

The other was the bridge of Nieulay:—

"and he ordered the Earl of Derby, his cousin, to take up a position on the said bridge of Nieulay, with plenty of men-at-arms and archers, in order that the French might not be able to pass, unless they passed through the marshes, which was not possible. Between the mount of Sangattes and the sea, on the other side opposite Calais, there was a high tower guarded by thirty-two English archers, to dispute the passage of the sand-hills by the French, from whose attacks it was strongly fortified with great double ditches."

The men at Tournay attack the tower and take it, after losing many of their number; but the marshals announce to Philip that they cannot go further without sacrificing a portion of his army. It was on this occasion that the King of the French took it into his head to send a message to the King of England :—

"Sire," said the envoys, "the King of France sends us hither to inform you that he hath come, and is now on the mount of Sangattes, for the purpose of encountering you; but he can neither see nor find out any way by which he may reach you, although he hath a great desire to raise the siege of his good town of Calais. He hath therefore made enquiries by his marshals how he might reach you, but he finds it a thing impossible. Therefore he would be glad if you would take counsel with those around you, and he with those around him, that so, by the assistance of these a place might be fixed for the combat: and to this end we are deputed to claim and require this at your hands."

A letter from the King of England to the Archbishop of York shews that this prince accepted the singular proposal of King Philip[z]; but that, after some parleying, during which the besieging army continued to fortify themselves more strongly in their camp and to defend their passages thereto, the King of France suddenly broke up his camp and dismissed his soldiers on the 2nd of August, 1347.

What precedes indicates that the military spirit was undergoing a modification in the West; and in this new path the Anglo-Normans had preceded us. At every turn in the fourteenth century the ancient chivalric spirit of the French comes into collision with the political bent of the Anglo-Normans, and with their national organization, already one in its nature, and powerful in consequence. The use of gunpowder in armies and sieges was another great blow to feudal chivalry. Individual energy, material force and headlong courage

---

[z] The narrative of Froissart is not in conformity with the king's letter; according to the chronicler, King Edward refused the *cartel* of Philip, saying that the latter had only to come and meet him in his camp.

would soon have to give way to the calculations, the forethought and the intelligence of the commander, seconded by troops accustomed to habits of obedience. Bertrand du Guesclin is the transitional figure between the knights of the twelfth and thirteenth, and the able captains of the fifteenth and sixteenth, centuries. It must be said that in France inferiority in warfare is never of any long duration; a nation which is warlike in its instincts learns still more from its defeats than even from its success. We have alluded to the distrust on the part of feudal France towards the lower classes, a distrust to which may be attributed the preference shewn in the army to the employment of foreign mercenaries over native troops, who, once dismissed, and having become accustomed to the use of arms and a life of danger, and numbering, moreover, one hundred to one, might have been able to combine against the feudal system and destroy it. Royalty, trammelled by the privileges of its vassals, could not directly call the population under arms; in order to get an army together, the king called upon his nobles, who responded to the appeal of their suzerain by bringing with them the men they were bound to furnish; these men composed a brilliant *gendarmerie* of picked troops, followed by *bidauds*, *valets*, *brigands*, forming rather a disorderly herd than a solid infantry. The king took into his pay, in order to fill up the blank thus left, Genoese or Brabançon archers, or those of the corporations of his *good towns*. The former, like all mercenary troops, were more inclined to pillage than to fight for a cause with which they had no concern; and the troops furnished by the great communes, turbulent in their nature, bound only to give a temporary service, and but ill disposed to go to any considerable distance from their homes, took advantage of the first reverse to return to their towns, abandoning the national cause, which indeed in their eyes had not yet an existence, since no true spirit of nationality could co-exist with the subdivisions of the feudal system. It was with such bad elements that Kings Philip de Valois and John had to struggle against the Gascon and English armies, already organized, compact, and regularly paid. They were beaten, as was natural. The unfortunate provinces of the north and west, exposed to the ravages of war, burnt and pillaged, were soon reduced to despair: men who had trembled before a coat of mail, while it appeared invincible, beholding the flower of the French chivalry destroyed by English archers and Welsh gallo-glasses, by simple foot-soldiers, that is to say, took up arms in their turn: indeed, what other course had they open to them? and formed the terrible companies of the Jacques. Those troops of brigands and dismissed soldiers, left to their own resources after a defeat, threw themselves upon the towns and the castles:—

"And there were always," says Froissart, "to be found poor brigands who would rob and pillage towns and castles, taking from them immense store of booty . . . . They could scent, as it were, a good town or castle at two days' journey; then some twenty or thirty of these brigands assembled, and they set forth and travelled day and night by secret paths, so that they entered the town or castle they had got scent of just as day was dawning, and set fire to a house or two. And those of the town fancied there were a thousand coats of mail come to burn their town: so they fled panic-stricken, and these brigands broke into houses and chests and caskets, and took whatever they found, and so went on their way loaded with booty . . . Amongst others was a brigand of Languedoc, who in this way marked out the strong castle of Combourne, which lies in the Limousin, in a country very difficult of approach. Thither he rode one night with a score and a-half of his companions, scaled and took the castle, and therein the lord of it, who was called the Vicomte de Combourne, and killed the whole of his household; the lord they imprisoned in his own castle, and kept him so long in durance, that he was fain to ransom himself for four and twenty thousand crowns, paid down. And the said brigand, further, kept the said castle to himself, furnishing it well, and made war upon the country. And afterwards, for his prowess, the King of France desired to have him near him, and bought his castle for twenty thousand crowns; and he became *huissier d'armes* to the King of France, and by him was held in great honour. The name of this brigand was Bacon. And he was always mounted on good steeds, and as well armed as an earl and as bravely attired, and so remained as long as he lived [a]."

Here we find the King of France making terms with a soldier of fortune, giving him a high position, and attaching him to his person; by thus acting the King made a great stride towards the defence of the national territory: he stepped beyond the limits of feudalism to summon to his aid chiefs sprung from the people. It was with these companies of soldiers, owning no country or allegiance, but brave and accustomed to the trade of arms, with these highway-men and freebooters, that Du Guesclin was about to reconquer, one by one, the strong places which had fallen into the hands of the English. Misfortune and despair had made soldiers of the people; even peasants held the country and attacked the castles.

While conquering a portion of the French provinces, the English had had to combat only the feudal nobles. After having taken their castles and domains, and finding that there was no *people* in arms, they left in their strong places only isolated and feeble garrisons—a few coats of mail supported by some archers; the English believed that the feudal nobility of France, however brave they might be, would not be able, without an army, to win

---

[a] Froissart, chap. cccxxiv., edit. Buchon.

back their castles. Great was the surprise of the English captains to find themselves, after an interval of a few years, assailed not only by a brilliant chivalry, but also by troops at once intrepid and disciplined in battle, obeying blindly the orders of their chief, having faith in his courage and his star, fighting with coolness, and possessing the tenacity, the patience, and the experience of veteran soldiers [b]. At the close of the fourteenth cen-

---

[b] No strong place could resist Du Guesclin; he knew how to carry his soldiers with him, and took nearly every town and castle by sudden attacks. He had discovered that the fortifications of his time could not resist an attack conducted without hesitation, promptly and vigorously. He delivered the assault by throwing a great number of brave, well-armed soldiers, provided with fascines and ladders, upon a given point; supported them by numerous crossbows and archers, under cover, and thus, forming a column of attack of devoted men, he lost but few of his men, by acting with vigour and promptitude. At the siege of Guingamp:—

"With trees and pieces of wood and branched bushes the bold assailants have filled the great moats; in two places or more are the planks already laid. To the gates comes Bertrand the bold, and loud he cried, 'Guesclin! now up with ye at once! for I must be lodged therein.' And they set up ladders, like good men and bold; whereby you might see mounting these undaunted burghers carrying on their heads great doors, and shutters, and shields, for fear of the stones which they flung on them from within. They who were inside were affrighted, and be sure they could not shew themselves at the battlements, because of the arrows sent against them. The castellain had gone up upon the donjon and watched the attack of those brave burghers, who were so hot in the assault that they cared nothing for death."

Du Guesclin employed no moving turrets, or other slow, costly, and difficult means of attack; he made use only of offensive engines; he employed the mine and sap, and ever with the activity, the promptitude, the abundance of resources, and carefulness in minor details, which characterize great captains.

He invests the donjon of Meulan:—

"The castellain was still within his tower: so strong was the tower that he had no fear. Well were they provided with bread and salted meat, and good wine enough to keep them yet fifteen months or more. . . . . . . Bertrand is gone to hold parley with the castellain, and he calls upon him to deliver up the tower, that it may be restored to the duke, whose deeds are worthy of so much praise. 'Ye,' says he, 'I will allow to depart in safety.' And the castellain answers, 'By the faith I owe St. Omer! If ye would lodge ye in this tower, methinks ye will have to take a high flight in air.' . . . . . . Bertran du Guesclin had the tower strongly assailed; but his assaults were of no effect; well were they provided to hold out for a long time. Then he made a mine, and the miners began their work, and he had them so guarded, that they could not be hurt; and the miners pushed on their work, and had the earth carried away so that those in the tower could not see them. So well did they make their mine, that they soon were able to come under

tury feudalism had played its part, military as well as political: its *prestige* was gone, and the troops which Charles VII. and Louis XI. possessed might be properly called regular armies.

If we have dwelt at some length on this question, it is because we have deemed it necessary to shew the several transformations through which the art of war has successively passed, in order to be able to make the different systems of defence, which were successively adopted from the tenth to the sixteenth century, better understood. It is needless to expatiate upon the arbitrary nature of the art of fortification, an art in which every other consideration should give way to the requirements of the defence; and yet such was the hold of the feudal traditions, that forms and arrangements were long preserved, so late indeed as the sixteenth century, which were nowise on a level with the new means of attack. It is especially to the fortifications of castles that this observation applies. Feudalism could not for a long time be induced to replace its high towers by low breastworks on an extended line; with it, the great donjon of stone, massive and close, was always the sign of strength and domination. And thus we find the castle suddenly passing in the sixteenth century from the fortification of the middle ages to the manorial pleasaunce.

The same thing does not occur in the towns: as the

---

the walls. From beneath the foundations they removed the earth, and had them sustained with many props, great and fair, strong and weighty, fixed thereunder. Then came the miners to Bertran, without stopping their work, and they said to Bertran, 'Sire, when you so shall desire it we will make this tower to fall.' 'Then so I will it at once,' thus speaks Bertran: 'for since those within will not obey, it is of right that they should die.' The miners have laid the fire within the mine, each in his own portion, the timber being first well smeared with bacon-fat; and the moment it was fired, as the song says, the high tower fell down like a crown from the brow." (*Chronicle of Bertran du Guesclin*, v. 3,956, and following verses.)

natural consequence of their disasters, the gendarmerie of France lost by little and little their ascendancy. Undisciplined, and ever placing feudal interests above the interests of the nation, they were reduced, during the wars of the fourteenth and fifteenth centuries, to the level of partisans, taking castles and towns by surprise, burning and pillaging them one day, and driven from them the next; holding now with one party and now with another, according as it suited the interest of the moment. But the corporations, of the *good towns*, who were unacquainted with warfare at the period of Edward the Third's conquests, had learned to fight; better disciplined, better armed, and braver than the gendarmerie, they formed, as early as the close of the fourteenth century, troops of whose tried valour the safeguard of important posts could be confided[c]. Towards the middle

[c] It was particularly during the fourteenth century that the corporations of archers and crossbow-men (*arbalétriers*) were regularly organized in the towns of the north. By an ordinance dated in the month of August, 1367, Charles V. establishes a company, or *connétablie*, of crossbow-men in the town of Laon. The King named Michauld de Laval constable for three years of this company. "Thereafter," says Article 1 of this ordinance, "the arbalestriers will elect every three years their constable by a majority of votes. Michauld de Laval, with the aid and counsel of five or six of the most experienced in the use of the crossbow, will choose the twenty-five arbalestriers who are to form the company. The arbalestriers will obey the constable in all that pertains to their duties, under a penalty of six sols." Article 2 decrees that "The king retains these arbalestriers *in his service;* and he places them under his safeguard." Then follow the articles which confer certain privileges on the company, such as their exemption from all imposts and taxes, with the exception of "the *aide* established for the ransom of King John."

The same prince establishes a company of twenty arbalétriers at Compiègne.

In the year 1359 the Corporation of Arbalétriers is established at Paris, to the number of two hundred: by an ordinance dated November 6, 1373, Charles V. fixes the number of these at eight hundred. These crossbow-men, who belonged to the middle class, and did not make their profession, were not allowed to quit their corporation, either to serve in the army or elsewhere, without the authorization of the provost of Paris and of the provost of the merchants. When these magistrates took the arbalétriers to any service beyond the *banlieue* of Paris, men and horses (for there were both horse and foot of them) were fed; each man received besides three sols, and the constable five sols, per diem: the whole at the cost of the city.

By letters patent of the 12th of June, 1411, Charles VI. ordained that a comof this century *cannon* had been used both in pitched battles and sieges [d]. This new means of destruction was destined to change, and did change before long, the whole conditions of the attack and defence of strong places. At the commencement of the fifteenth century, the new artillery was still of little importance, but it assumes a great development towards the middle of this century:—

"In France," says the illustrious author already quoted [e], "the war of independence against the English had re-awakened the warlike genius of the nation, and not only the heroic Joan of Arc occupied herself with the direction of the artillery [f], but two eminent men sprung from the ranks of the people, the brothers Bureau, gave their whole attention to the improvement of great guns and the conduct of sieges. They began by employing, although at first in small numbers, iron instead of

---

pany of archers, composed of one hundred and twenty men, should be established at Paris; that these one hundred and twenty should be chosen amongst the other archers then existing; and that this company should be specially charged to guard the person of the king and with the defence of the city of Paris.

Charles VII., by letters patent of the 22nd of April, 1448, instituted the body of free-archers (*francs-archers*), to serve in time of war. For the formation of this privileged corps there were chosen in each parish, robust and skilful men amongst the wealthier inhabitants, because those free-archers were obliged to equip themselves at their own expense, or, in default, at the cost of the parish. The ratio of the contingent was about one man to fifty hearths. (*Recherches Hist. sur les Corpor. des Archers, des Arbalétriers et des Arquébusiers,* par Victor Fouque, 1852, Paris.)

[d] The English army had cannon at the battle of Crécy. From the year 1326, the city of Florence had cannons manufactured both of iron and metal. (*Bibl. de l'Ecole des Chartes,*—"Library of the School of Charts,"—vol. vi. p. 50.) In 1339, two knights, the sires de Cardilhac and de Bieule, receive from the master of the arbalétriers of the town of Cambray, "ten cannons, five of iron and five of metal" (probably of *wrought* iron and *cast* metal), "the which were all made by the commandment of the said master of the arbalestriers by our hands and by our people, and which are for the guard and defence of the town of Cambray. (*Original parchment amongst the sealed title-deeds of Clairambault,* vol. xxv., fol. 1,825; *Bibl. de l'Ecole des Chartes,* vol. vi. p. 51.) "For saltpetre and sulphur . . . purchased for the cannons at Cambray, eleven livres four sols and three deniers *tournois.*" (Ibid.)

[e] The Past and Future of Artillery, by L. Napoleon Bonaparte, vol. ii. p. 96.

[f] Deposition of the Duc d'Alençon, Michelet's Hist. of France, vol. v. p. 99.

stone balls[g], and by this means a projectile at the same weight occupying a smaller volume, a greater quantity of movement could be imparted to it, because the piece having a less calibre, offered a greater resistance to the explosion of the powder.

"This harder ball was not liable to fracture, and was able to penetrate masonry; there was besides an advantage in increasing its velocity by diminishing its mass; the bombards were lighter, although their effect was rendered more dangerous.

"Instead of erecting bastilles all round the town[h], the besiegers established before the great fortresses a park surrounded by an entrenchment, beyond the reach of cannon. From this point they conducted one or two branches of the trenches towards the points were they established their batteries[i] . . . We have arrived at the moment when trenches were employed as a means of approach concurrently with covered ways of timber[k] . . . . .

---

[g] The *trébuchets*, *pierriers*, and *mangonels* threw stone balls: it was natural, when altering the mode of projection, to preserve the projectile.

[h] See the siege of Orleans in 1428. We shall return to the works executed by the English to batter the walls and blockade the town.

[i] At the siege of Caen, in 1450: "Thereafter they began on the side of Monseigneur the Constable to make covered and uncovered approaches, whereof le Bourgeois undertook the conduct of one and messire Jacques de Chabannes of the other; but that of le Bourgeois was the first at the wall, and the other arrived afterwards, and the wall was mined at that place. Insomuch that the town would have been taken by assault, but for the king, who would not have it, nor would he send any bombards on that side, for fear the Bretons should make an assault." (*Hist. d'Artus III., Duc de Bretaigne et Connest. de France, de nouveau mise en lumière*, par T. Godefroy, 1622.)

At the siege of Orleans, 1429: "On Thursday, the third day of March, in the morning, the French sallied forth against the English, who were making at that time a trench to go under cover from their boulevard of Croix-Boissee to Saint Ladre d'Orléans, in order that the French could not see them, or do them hurt by means of cannons and bombards. This sally did great damage to the English, for nine of them were taken prisoners; and besides there was killed maître Jean by a double culverin of five (*coulevrine cinq à deux coups*)." (*Hist. et discours du siége qui fut mis devant la ville d'Orléans.*—Orleans, 1611.)

[k] We cannot, however, admit that trenches were never employed as a means of approach before the artillery of great guns came into use. Philip Augustus, at the siege of Chateau-Gaillard, had regular trenches made in order to close with the works which he wished to attack first; those trenches conducted his troops and his *chats* or timber galleries up to the counterscarp of the moat. Going still further back in the history of sieges, we find trenches indicated as means of approach

To the brothers Bureau belongs the honour of having been the first to make a judicious use of artillery in sieges: all obstacles fell before them, the walls struck by their balls were incapable of resistance, and flew into fragments. The towns defended by the English, and which, at the time of their invasion, they had taken months to besiege, were carried in as many weeks. They had spent four months in besieging Harfleur in 1440; eight months in besieging Rouen in 1418; ten months in taking Cherbourg in 1418; whilst in 1450, the conquest of the whole

---

to strong places. In the Πολιορκητικὰ of Hero of Constantinople, written in the sixth century, and compiled in the tenth, we read this curious passage. A place situate on the top of the hill is to be attacked: "There is yet another means of preservation from the masses rolled from above. We must, beginning at the foot of the hill, dig oblique ditches, directing them upwards against a certain part of the walls: these ditches must have a depth of about five feet, and a wall built vertically on the left side of the same ditches, in such wise that the masses rolled down from above should strike against this wall, which serves as a rampart and shield to the assailants. The labourers should fortify the portion of the ditch already dug, in the following manner: they should take stakes of wood some three ells long, or trunks of young trees, and sharpen them at the lower extremity; these they should drive into the ground, so as to offer resistance, on the left of the before-named wall which rises above the earth thrown from the ditch, and give them an oblique position as regards the slope of the hill; they should then place planks externally against those stakes, and fasten all round branches of trees woven together (fascines); finally, throwing up on this side all the materials obtained in excavating, they should prepare straight roads for the ascent of the *tortoises*. These tortoises, seen in front, should be those they call *spurs*, that is to say, coming to an acute angle in front, and based on a triangle or pentagon, and constructed on this large base they narrow gradually upwards to the ridge which forms the top of the machine, so that they resemble in front the prows of ships set on the ground and resting one against another. They must be small and numerous, so that they can be prepared quickly and easily, and be carried by a few men. They should have at their base spikes of wood each a foot long, and iron nails instead of wheels, to the end that when placed on the ground they should be fixed, and not be carried backwards by a shock. Further, each of them should have at its front a piece of oblique wood, like that which chariots have at their fore-part, to arrest and keep it in its place when it might otherwise slide down the slope of the hill, especially when those who are pushing it forward up the slope are tired and want to rest a moment. Thus one of three things will come to pass; either the projectiles hurled from the top of the hill, falling into the ditch, will be turned into another direction, or, striking against the stakes obliquely inclined, they will be stopped in their course, or else impinging on the spur of the tortoise, they will be thrown either on one side or the other, and the intermediate space will be sheltered from their blows. . . . . ." (*Acad. des Inscript. et Belles-lettres: Mém. présentés par divers savants*, 1re *série*, tom. iv.; *Morceaux du texte grec inédit des* Πολιορκητικὰ *d'Héron, de Constantinople, publ. d'après les manusc. d'Oxford*, trad. de M. Th. Henri Martin.)

of Normandy, which it required sixty sieges to accomplish, was effected by Charles VII. in one year and six days[1].

"The moral effect produced by the artillery of great guns had become so great, that it was only necessary for them to be brought on the ground, to make a town surrender.

". . . . Let it be said, then, to the honour of the arm, that it is as much to the progress of artillery as to the heroism of Joan of Arc that France is indebted for having been enabled to throw off the yoke of the foreigner from 1428 to 1450. For the dread which the great had of the people and the dissensions of the nobles would perhaps have led to the ruin of France if the artillery, ably conducted, had not appeared to give a new strength to the royal power, and to furnish it with the means of driving out the enemies of France, and of destroying the castles of those feudal lords who did not acknowledge a fatherland.

"This period of history marks a new era. The English have been vanquished by the new guns, and the King, who has won back his throne by plebeian hands, finds himself for the first time at the head of troops who belong to himself alone. Charles VII., who at a former period borrowed from the towns the cannon wherewith to make his sieges, now possesses an artillery numerous enough to carry out attacks upon several places at the same time; a fact which justly excites the admiration of his contemporaries. By the creation of companies of ordnance, and the establishment of free-archers, the King acquires a force of cavalry and infantry independent of the nobility. . . ."

---

[1] ". . . . . And siege was laid to Cherbourg. And my said lord encamped on one side, and Monseigneur Clermont on the other. And the Admiral de Coitivi and the Marshal and Joachim were on the other side, over against a gate. And the siege lasted a full month, and there were broken and injured nine or ten bombards, great and small. And the English came there by sea, amongst others a great ship called the ship Henry, and the mortality set in a little, and Monseigneur had much to suffer, for it was he who had the whole charge. Then he placed four bombards on the seaside, in the sands, when the tide had run out. And when the tide flowed in, all the bombards were covered, *mantels* and all, and they were all loaded, and so well covered up, that as soon as the tide had run out again they had only to set the matches to them, to fire them as well as if they had been on dry land." (*Hist. d'Artus III.*, p. 149.)

The use of great guns in sieges would have as its first result the doing away everywhere with hoards and brattishes of wood, and the substitution for them of machicolations and pierced parapets of stone, carried upon stone corbels projecting before the face of the walls. For the first, cannon appear to have been often used, not only for hurling round stones as bombs, like the engines which worked by counterpoise, but likewise for throwing small barrels containing an inflammable and detonating composition, such as the Greek fire described by Joinville, and known to the Arabs from the twelfth century. At the end of the fourteenth and beginning of the fifteenth century the artillery have already begun to throw balls, of stone, lead, or iron, horizontally; they no longer confine the attack to the battlements and upper defences of the walls, but effect breaches at the base: the true siege battery is established. At the siege of Orleans, in 1428, the English threw into the town, with their bombards, a large number of stone projectiles, which pass over the walls and crush the roofs of the houses. But, on the side of the French, we find an artillery who fire point-blank, causing great losses to the besiegers; the Earl of Salisbury is killed by a cannon-ball, while observing the town through one of the windows of a turret[m]. It is a man sprung from the people, a Lor-

m "During the festival and service of Christmas they fired, on one side and the other, very horribly and incessantly, from bombards and cannons; but specially there was one who did great damage, a culveriner, native of Lorraine, being then one of the garrison at Orleans, named *Maitre Jean*, who was reputed the best master of that art then known, and well he proved it; for he had a great culverin which he fired many times, being then within the piers of the bridge, near to the boulevard de la Belle-Croix, so that he wounded and killed many of the English." (*Hist. et Discours, &c. du Siège d'Orleans.*)

"This same day (the last but one of the month of February, 1429) the bombard of the city, then fixed near to the Chesnau postern, to fire against the turrets, shot so terribly against them, that it threw down a great piece of the wall." (*Ibid.*)

"The French attacked the said castle of Harecourt with an engine, and with the rainer called Maître Jean, who directs the artillery of the town.

To besiege the town, the English still follow the ancient plan of wooden bastilles and boulevards; they end by being themselves, in their turn, besieged by the men of Orleans, and they lose their bastilles one after the other, which are destroyed by the fire of the French artillery. Vigorously attacked, they are obliged to raise the siege, abondoning part of their *matériel;* for the siege artillery, like all the engines then employed, had the inconvenience of being difficult of transport; nor was it until under the reigns of Charles VII. and Louis XI. that siege-pieces, as well as those used in the field, were mounted on wheels; bombards, (great pieces, somewhat like mortars, used for throwing stone bullets of large diameter,) however, continued to be employed until during the latter years of the sixteenth century. We give (fig. 79) the representation of a double siege-gun provided with its *mantelet* of wood, intended to protect the piece and the gunners serving it against projectiles; (fig. 80), the drawing of a double cannon, but with chambers fitting into the breech, and containing the charge of powder and the ball [a]. Besides the

---

first shot they fired against it they pierced through and through the wall of the lower court-yard, which is very far, to the *equipolent* of the castle, which is of great strength." (*Alain Chartier*, p. 162, Ann. 1449.)

[n] Copied after vignettes of the manusc. of Froissart, 15th cent., Imperial Library (of Paris), No. 8,320, vol. i. The cannons (fig. 80) are shewn in the vignettes entitled "How the King of England laid siege to the city of Rains" (Rheims) .... "How the town of Duras was besieged and taken by assault by the French." These guns were at first made with bands of wrought iron joined together like the staves of a cask, and encircled by other cylindrical bands of iron. There still exists in the court-yard of the arsenal of Bâle a fine piece of ordnance so made, very carefully wrought; its length is 8 feet 11 inches, and it takes a ball of stone 12$\frac{80}{100}$ inches in diameter. The breech is forged in a single piece and contains a chamber of a less calibre than the bore. When the pieces were of small calibre, they were either wrought or cast, of iron or copper.

piece are other chambers of a similar kind, of which one, C, is furnished with a handle (see at the Artillery Museum (of Paris) guns furnished with this kind of chamber); (fig. 82), the drawing of a boxed cannon mounted on a carriage, with notched quadrant, for pointing the piece. The balls of this last cannon are of

Fig. 79. A Double Cannon with the Wooden Shield or Mantelet.
Fig. 80. A Double Cannon with the Chamber for Powder.
C. A Chamber with a handle.
Fig. 81. A Cannon mounted on a Carriage with a Quadrant.

stone, whilst those of the double cannons are of metal. The piece was fired by applying a metal bar made red-hot in the furnace to the powder contained in the chamber. The ranging of these pieces in battery, the loading of them,—especially when after each discharge the boxes or chambers had to be changed,—the means required for applying the fire, all this required much time. At the commencement of the fifteenth century the cannons of large calibre used in sieges were not sufficiently numerous; the great difficulty attending their transport did not allow of their being discharged with sufficient frequency to produce prompt and decisive effects in the attack of strong places. It was necessary to have, in order to keep the defenders from the battlements, numerous bodies of archers and crossbow-men; of archers more particularly, who had, as we have al-

Fig. 82. An Archer with his Sheaf of Arrows.

ready mentioned, a great superiority over the crossbows by reason of the rapidity of their fire. Each archer (fig. 82) was furnished with a leathern bag containing two or three dozen arrows. While in action, he laid his bag on the ground, open, and kept under his left foot

Fig. 83. An Archer firing downwards.

some arrows, the points towards his left; thus, without seeing, he could feel them and could take them up one by one without losing sight of his aim (an important

Fig. 84. A Crossbow-man with his Shield on his back, taking aim. From a MS. of Froissart.

point for a marksman). A good archer could shoot ten arrows per minute; whilst a crossbow-man in the same space of time could shoot only two bolts (figs. 84, 85). As he was obliged to fit the gaffle (fig. 86) or handle to his arm after every shot, in order to bend his bow, he not only lost much time, but he also lost sight of the movements of the enemy, and was obliged, every time his crossbow was strung, to seek his object out and take fresh aim °.

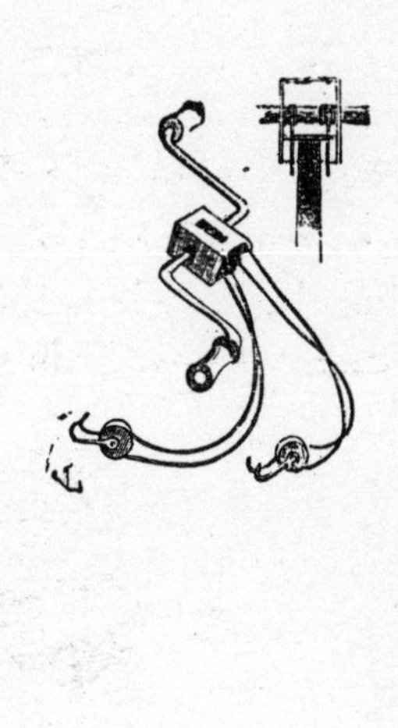

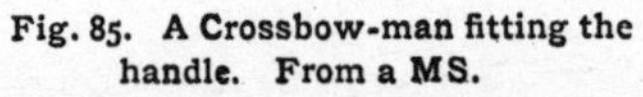

Fig. 85. A Crossbow-man fitting the handle. From a MS.

Fig. 86. The Cranequin, or Handle of the Crossbow.

When the new artillery was sufficiently well mounted, and could be used in such force as to enable the besiegers to breach the walls from a distance, the ancient system of defence became so inferior to the means of attack that it was found necessary to subject it to im-

° These figures are taken from the manuscript of Froissart already quoted. One of the crossbow-men (fig. 84) is what was termed *pavaisé*, that is to say, he bears on his back a large *pavois*, or shield, attached to a thong; whilst in the act of turning to prepare his piece, he was thus sheltered from the enemy's fire. The iron ring fixed to the bottom of the crossbow served as a stirrup for the foot when using the gaffle to bend the bow (fig. 85).

portant modifications. The ancient towers, covered for the most part with roofs of small diameter, and vaulted commonly in a slight way, were not adapted for receiving cannon; by removing the roofs and forming platforms instead (an operation frequently performed in the middle of the fifteenth century), it was possible to place one or two pieces at the top; but these could inflict no great damage on the assailants, as their plunging fire could strike only at one point. Their position had to be constantly altered, in order to follow the movements of the attack, and their recoil frequently shook the walls to such a degree as to make them more dangerous to the besieged than to the besiegers. On the curtain-walls, the parapets, which were only some two yards in width at the utmost, could not receive cannon; to remedy this, the ground was filled up on the inside to the level of the parapet, where guns were to be fixed and placed in battery; but still the curtains were so high that the fire was oblique and did not produce a great effect. While still continuing, therefore, to place artillery on the summits of the defences, embrasures were opened, wherever practicable, in the lower stories of the towers, on a level with the top of the counter-scarp of the ditches, in order to obtain a horizontal fire, to be able to send projectiles *en ricochet*, and to force the assailants to begin their approaches at a great distance and to sink their trenches to a considerable depth. Under Charles VII., in fact, many castles and towns had been successfully carried by sudden attacks. Guns were brought up at once to the walls, without any cover or trench, and before the besieged had time to place the few bombards and *ribaudequins* [p], with which the towers were armed, in battery, the breach was

[p] *Ribaudequin*, a kind of huge crossbow, fifteen feet in length, for throwing darts five feet long.

made and the town taken. But all towers were not equally susceptible of the modifications required before making use of artillery in defence; some were of an internal diameter which did not admit of receiving pieces of ordnance, nor could they always be introduced through the winding corridors and staircases of these buildings; and even when fixed, after two or three shots, there was a risk of being smothered by the smoke, which had no means of escaping. In the places which were fortified towards the middle of the fifteenth century, we can perceive that the new artillery has begun to engage the attention of the architects; they do not as yet abandon the ancient system of curtains with flanking towers, a system consecrated by long custom; but they modify it in the details, they extend the line of external defences, and no longer place cannon at the summits of their towers. Reserving these crest-works for close defence, they furnish the lower parts of the fortifications with artillery.

The study of this transition is very interesting: it is rapid because of the improvements introduced into the attack of strong places, which forced the constructors to modify, from day to day, their defensive measures. We possess few complete military structures which have preserved intact the arrangements made in the time of Charles VII. for resisting an artillery, already of formidable strength. There is one, however, which we shall give here, as well on account of its state of preservation, as because it was erected as a whole in its present form, and because its system of defence is carried out methodically in all its parts: this is the castle of Bonaguil. Situate at a distance of a few leagues from Villeneuve d'Agen, this castle is built upon a spur or bluff commanding a defile; its site is that of all feudal castles of any importance; surrounded by precipices, it is only accessible from one side (fig. 87[q]), at A. A drawbridge gives access to an advanced work which the constructors have been at great pains to flank effectually. At O is a *place-d'armes*, and at R were probably the stables. A wide ditch separates this advanced work from the castle, which is entered by a second draw-

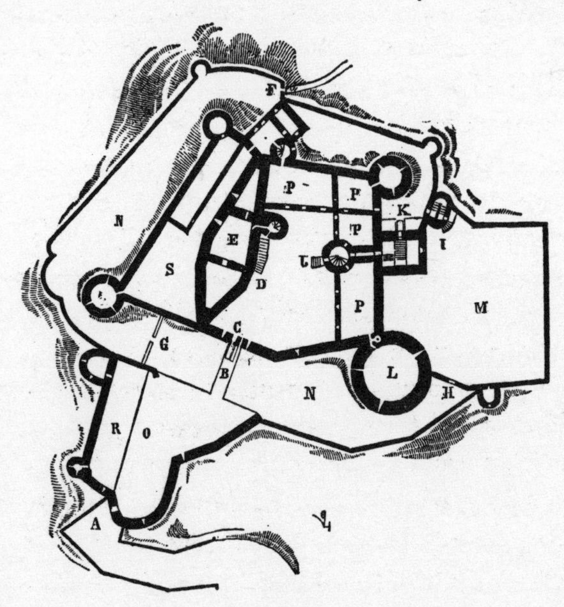

Fig. 87. Plan of the Castle of Bonaguil.

A. Outer Drawbridge.
B. Second Drawbridge.
C. Gate.
D. Staircase.
E, F, G, H. Gates.
J. Winding Stairs.
K. Drawbridge.
L. Round Tower.
M. Platform.
N. The Moat.
O. Parade-ground.
P P. The Barracks.
R. The Stables.
S. Outwork.

bridge, B, and a gateway with postern, C. A donjon, E, of unusual form, commands the outside and the outwork, O. At P are placed the dwelling apartments, reached by a fine spiral staircase, J. D is the staircase

[q] This plan is drawn to the scale of $\frac{1}{1000}$, which is the scale of the plans of the castles of Montargis and Coucy, already given.

leading to the entrance (which is on a higher level than the ground) of the donjon, E. At S is a work separated from the castle by the donjon. When the drawbridges were raised, the castle could be entered only by passing through the gate, F, pierced in the wall of counterguard; by following the bed of the moat, N; then passing through a second gate, G, in the centre of a traverse, and a third gate, H, opening upon a fine platform, M; taking the staircase, I, and passing over a smaller drawbridge, K. There we come upon a wide and handsome staircase, communicating with the internal staircase, J, only by a dark and narrow corridor, loopholed on both sides. The great staircase stops at the level of the ground-floor (at a height of some feet above the ground) of the internal court; the upper portion forms a great square tower. We here find all the precautions which were taken in the ancient feudal castles to mask the entrances and render them difficult of access. But arrangements, then quite novel in their character, were made for modifying the ancient defensive system; firstly, the advanced work, with the platform, M, form considerable salients or projecting parts, which command the outside to some distance; then, at the level of the counterscarp of the ditches, or that of the top of the walls of counterguard, embrasures are pierced in the ground-story of the curtains and towers, to receive cannon; and the towers are almost detached to flank the curtains more effectually. If we may judge from the doorways opening into the towers, the pieces placed in battery could not be of large calibre. All the crest-works are provided with battlements and machicolations; but the merlons of the parapets, still standing, are pierced with loops of such an arrangement as to indicate clearly the use of ordnance. Subjoined (fig. 88) is a view of this castle taken from the side of the entrance[r]. We see in this view that these embrasures which are intended for artillery are pierced in the lower stories of the buildings, that they follow the declivities of the

Fig. 88. Bird's-eye view of the Castle of Bonaguil.

ground, or are made to command the works in front. As for the crest-works of the towers, they are the same as those adopted in the fourteenth century. The transi-

[r] We have only added in this view the timber-work, which no longer exists; the masonry remains almost intact.

tion is thus evident, and might be summed up in the following formula:—*To command the outside parts at a distance and the approaches, by a horizontal fire of artillery, and to provide against escalade by works of a great elevation with crest-works, according to the ancient system for close defence.*

The embrasures for cannon in the castle of Bonaguil are thus constructed. A gives the plan of one of them; B, the internal opening; C, the port-hole on the outside. There is only room for the passage of the ball, with a sighting loop over (fig. 89).

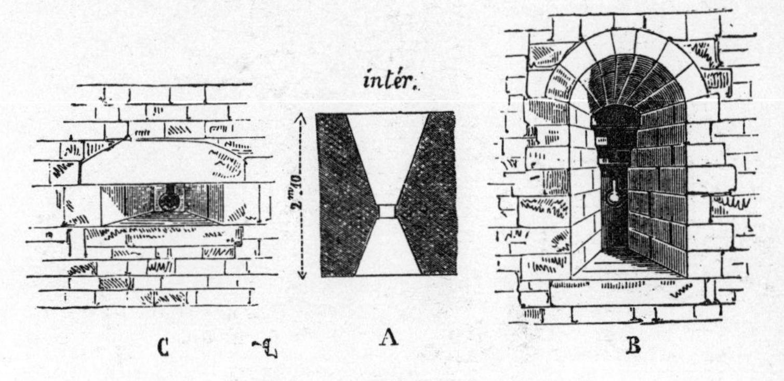

Fig. 89. Embrasure of the Castle of Bonaguil.

C. Exterior. A. Plan. B. Interior.

These different improvements, however, were still not adequate to meet the means of attack. The divergent fire of a few pieces at the foot of the towers and curtains rendered their effect almost insensible upon siege batteries composed of several pieces brought together at a single point. While the defenders were sending one ball, they received twenty; the works of defence were riddled with shot concentrated upon a single point, and fell in ruins before their cannon could inflict any sensible damage upon the besiegers. When this insufficiency on the part of the ancient system of fortification was clearly proved, engineers acted as the men of those times always did,—they applied the remedy where they saw the evil; the ancient system was preserved, but the constructors endeavoured to give their works a greater force of resistance. They began, first, by modifying the towers, which they built of less height and of a much greater diameter, giving them more and more external projection; abandoning the ancient system of isolated defences, they left the towers open on the inside in order to be able to introduce cannon with greater facility; they pierced them with more numerous lateral embrasures, below the level of the crest of the ditches, and enfilading these along their whole length; the lower stories were reserved for flanking the curtains at the moment the enemy appeared upon the ditch, and the upper stories for commanding the outlying parts as far as possible.

The fortifications of the town of Langres are a very interesting study, viewed in relation to the modifications introduced in the defence of places during the fifteenth and sixteenth centuries (fig. 90 [s]). Langres is a Roman town; the portion, A, of the town was added at the commencement of the sixteenth century, to the ancient Roman enclosure, of which there remains a gateway still in a good state of preservation; the walls of Langres, after having undergone successive modifications, were almost entirely rebuilt under Louis XI. and Francis I., and further strengthened with defences erected according to the system adopted in the sixteenth century and at the beginning of the seventeenth. It was the introduction of ordnance which led to the erection of the towers, C, in order to flank the curtains by means of two parallel walls terminating in a semicircle. The town of Langres

[s] This plan is taken from the *Topographie de la Gaule*, Frankfort edition, 1655. The greater part of these fortifications still exist.

is built upon a plateau which commands the course of the Marne and all the surrounding country: on one

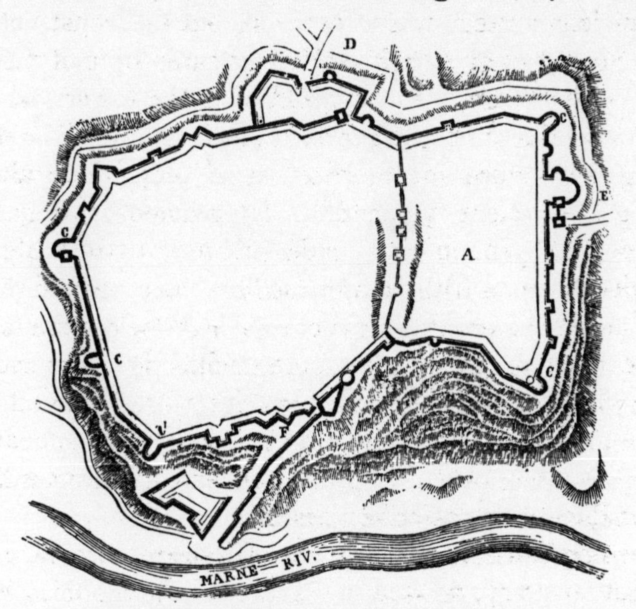

Fig. 90. Plan of the Walls of the Town of Langres.
A. Part added. B. Cross Wall. CC. Flanking Towers. D. Outwork.
E. Second Gate. F. Third Gate.

side only, D, it can be approached from the level, and on this side we find that an advanced work had been thrown up in the sixteenth century [t]. At E was a second gateway, well defended by a massive round tower (or boulevard) with two covered batteries placed within two chambers, the vaulting of which rests upon a central cylindrical pillar; in another tower on the opposite side is a spiral ramp, or inclined way, for the purpose of getting up cannon to the platform which crowns the great tower; at F a third gate opening on the Marne, protected by earth-works dating from the close of the sixteenth century. We give (fig. 91) the plan of the

[t] The advanced work shewn upon this plan has been replaced by an important modern defence, which spans the road from Dijon.

ground-floor of the great tower or boulevard defending the gateway, E; (fig. 92) the plan of the first story of

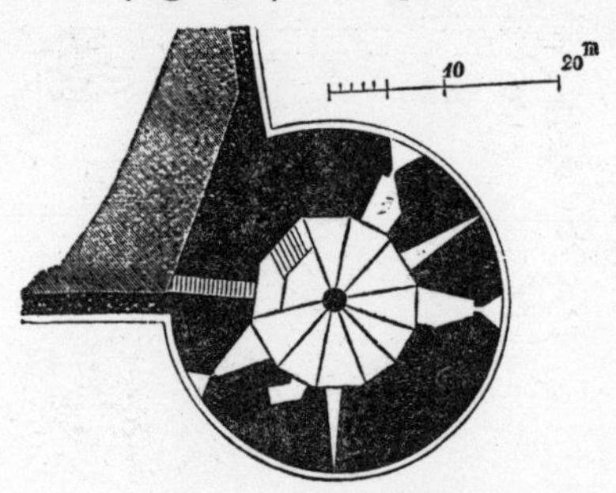

Fig. 91. Ground-plan of the Great Tower, Langres.

the same. If we examine the second plan, we find that the embrasures for cannon are so arranged as to enfilade

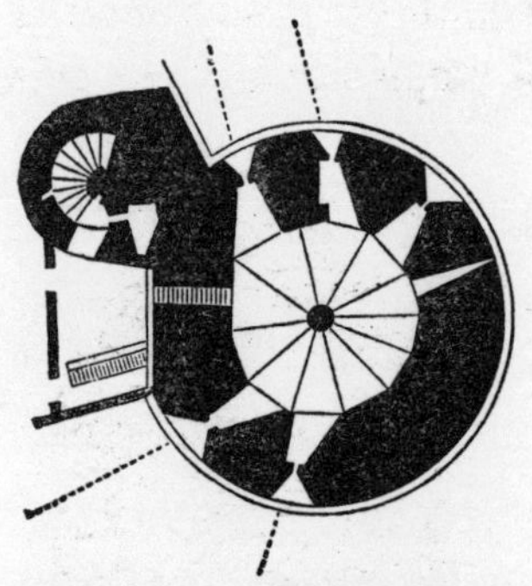

Fig. 92. Plan of the First Story.

the curtains. Fig. 93 gives the section of this boulevard, at the summit of which a *barbette* battery could be established. We subjoin (fig. 94) the plan of one of the towers of the fortifications at Langres, the erection of which, as well as that of the boulevard, dates from the beginning of the sixteenth century. This tower, built upon a rapid incline, is a true bastion, capable of receiving on each story five cannon. We descend successively four flights of steps from the point C, opening into the town to the point E. The embrasures, E, F, G, are stepped to follow the inclination of the site, so as to be all placed at the same elevation above the ground out-

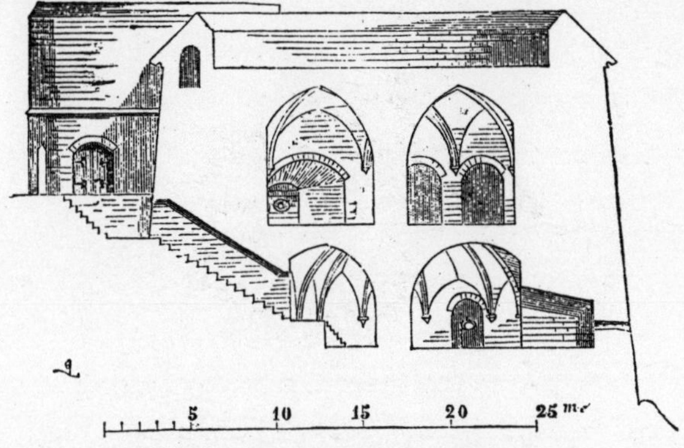

Fig. 93. Section of the Tower, Langres.

side. Cannon could be easily moved up or down the several flights of steps, which are wide and not steep;

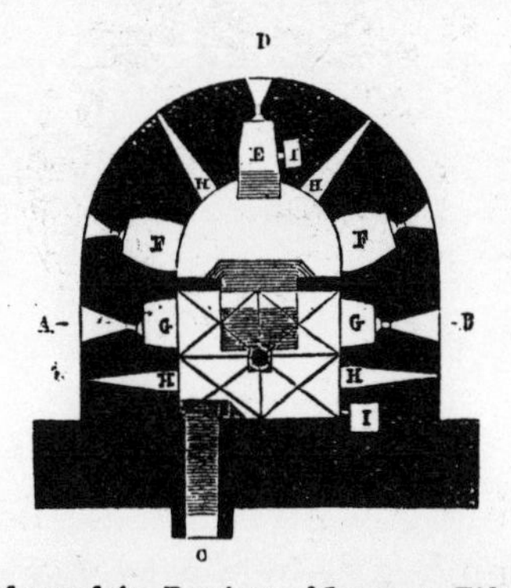

Fig. 94. Plan of one of the Bastions of Langres, Fifteenth Century.

A B. Line of the Section, fig. 96. C D. Line of the Section, fig. 95.
E. Highest level towards the Town. F F, G G. Embrasures.
H H. Vent-holes. I I. Small Closets.

the walls are thick (about 21 feet) in order to resist the artillery of the besiegers. The first bay (or division), the walls of which are parallel, is sustained by four vaults resting upon a single column; an archway connecting the two ends of a partition-wall divides the first bay from the second, which has a quarter-sphere vault (see the longitudinal section (fig. 95) on the line

Fig. 95. Section of the Bastion on the line C D of the Plan, fig. 94.

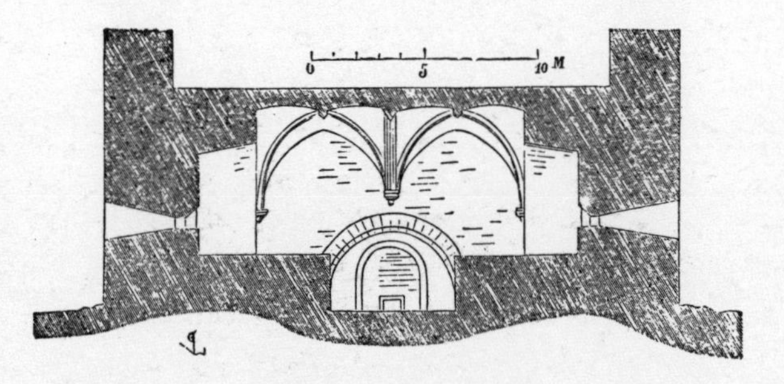

Fig. 96. Section of the Bastion on the line A B of the Plan, fig. 94.

C D, and the transverse section (fig. 96) on the line A B of the plan). The embrasures, F, G (fig. 94), were closed by shutters on the inside. Vent-holes, H, allowed the smoke to escape from the interior of the chamber. Two small closets, I, were probably powder-magazines. This tower was crowned originally by a platform and a pierced parapet, behind which other pieces of ordnance might be placed, and harquebus-men : these upper portions have been long altered. The barbette battery overtops the crest of the parapet of the adjoining curtains by about a yard, and offers thus another example of the influence of mediæval traditions ; according to which, the towers should always command the curtains [u]. This uncertainty and vacillation during the first period of the use of artillery led to the adoption of a great variety of different arrangements, all of which it is impossible we can give. But it may be well to observe that the system of fortification so well established from 1300 to 1400, and so methodically combined, was suddenly deranged by the intervention of artillery in sieges, and that the new course of experiments commences with the latter of these dates, to close only as late as the seventeenth century. Such was the hold of the feudal traditions on military engineers that they could not break suddenly with them, but continued subject to their influence long after the inconveniences attached to the mediæval system of fortification, as opposed to artillery, had been discovered. It is thus we find, even as late as the sixteenth century, machicolations employed concurrently with covered batteries, although machicolations were useless as a defence against cannon. So, from the time of Charles VIII. to that of Francis I., the towns and castles cannot hold out against armies provided with artillery, and history during this period offers no examples of those prolonged sieges, which are so frequent during the twelfth, thirteenth, and fourteenth centuries. They did the best they could to adapt the ancient fortifications to the new modes of attack and defence; either by allowing the ancient walls to remain behind new works, or by doing away with the weak portions of the former, as at Langres, in order to replace them by massive round or square towers furnished with artillery. At the end of the fifteenth

---

[u] This tower is at present called *La tour du Marché* (Market-tower). We give the only story which is preserved, the lower one. The plan is drawn to a scale of $\frac{7}{1000}$ full size.

century, engineers appear to have sought to cover their pieces of ordnance; they place them on the ground-story of towers in covered batteries, reserving the crest-works of towers and curtains for archers, crossbows, and harquebus-men. There are still in existence a large number of towers having this arrangement; without speaking of that at Langres, which we have given (figs. 94, 95, 96), but the crest-works of which, now destroyed, cannot be cited as an example, here is a square tower attached to the defence (of great antiquity) of Puy-Saint-Front at Périgueux, and which was reconstructed to take the cannon on the ground-floor level[x], intended to command the river, the river-bank, and one of the two curtains. The ground-floor, which is but small, of this tower (fig. 97) is pierced with four embrasures intended for as many small pieces of artillery, without reckoning a loophole at the salient angle, on the side opposite the river. Two cannons only (having to be moved from place to place according to the requirements of the defence) could be contained in this low battery, which was covered by a massive barrel-vault in masonry, and proof against solid projectiles thrown as bombs. The embrasures for cannon (fig. 98) are pierced horizontally, leaving just space enough for the passage of the ball; a horizontal slit, over the embrasures, facilitates the pointing of the piece and allows the smoke to escape. A straight stair leads to the first story, which is pierced only with

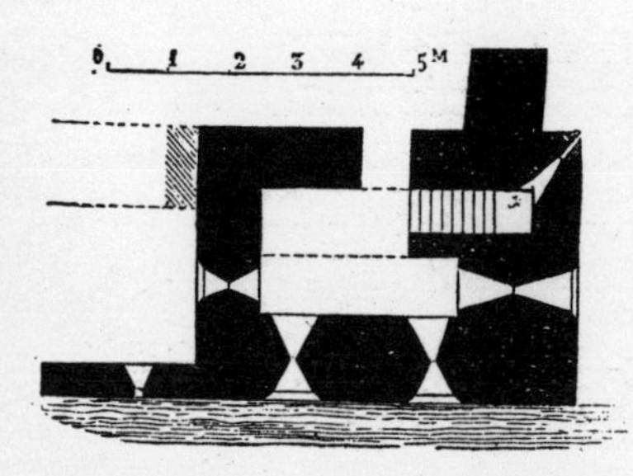

Fig. 97. Plan of a Tower at Perigueux.

[x] The adjoining curtains belong to the thirteenth century.

loops for crossbows and harquebuses, and the crest-work consists of a machicolation with continuous pa-

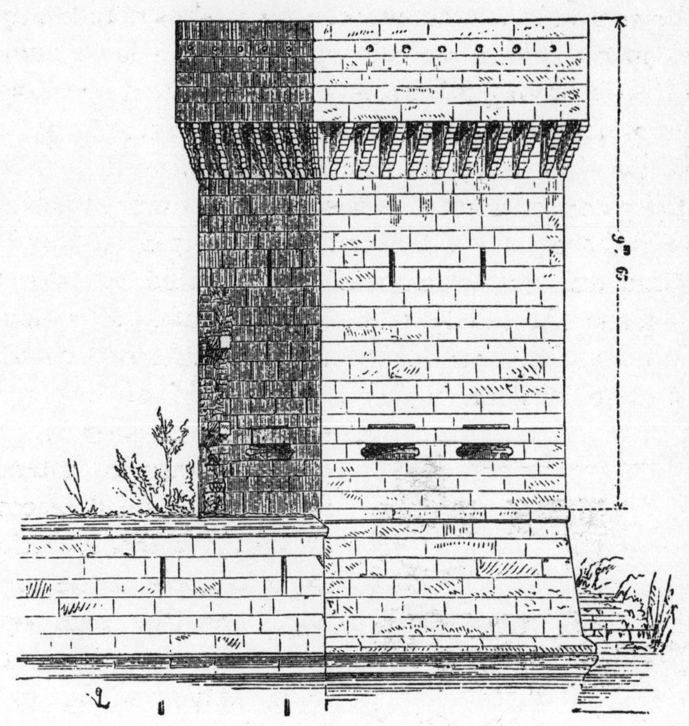

Fig. 98. View of the Tower at Perigueux.

rapet without crenelles, but pierced with round holes to take the barrels of small culverins or hand-harquebuses. As a defence this work was feeble, and it was easy for an enemy to place himself in such a position as to be out of the reach of its fire. It was soon found, firstly, that those covered batteries, fixed within small chambers, and the embrasures for which covered only an acute angle, could not dismount siege-batteries, and inflicted no serious damage on the assailants; and, secondly, that it was necessary to adopt a general system of flanked defences appropriate to the new mode of attack. Amongst the attempts which were made at the end of the fifteenth century, and the beginning of the sixteenth, to place the defence of strong places on a level with the attack, we must not omit to mention the fine fortress of Schaffhausen, which offers a perfect system of works very remarkable for the period, and still at the present day in a complete state of preservation. In order, however, to estimate justly the importance of this fortification, it is necessary to take its site into account. Issuing from the Lake of Constance, the Rhine flows past Stein, westward; and when it reaches Schaffhausen makes a sudden bend to the south as far as Kaiserstuhl. This bend is caused by some high rocky hills which presented an obstacle to the flow of the river, forcing it to change its course. Stein, Schaffhausen and Kaiserstuhl form the three angles of an equilateral triangle, of which Schaffhausen is the apex. It was, therefore, highly important to fortify this advanced point, the frontier of a state, and more especially as the left bank of the river (that which is within the triangle) is commanded by the hills on the right bank to which we have alluded as offering an insurmountable barrier to the course of the river. In case of invasion the enemy would not fail to occupy the two sides of the triangle and attempt the passage of the river at the point where it makes the bend. So much being premised, what the Swiss did was this: they constructed a bridge uniting the two banks of the Rhine, and the two parts into which the town of Schaffhausen is divided; and on the right bank they planted a vast fortress commanding the river, connecting the citadel so built with the Rhine by two walls and some towers. Those two walls form a vast triangle, a kind of tête-de-pont, commanded by the fortress. We give (fig. 99) the general aspect of this fortification, which we will proceed to study in its details. The citadel, or rather the great boulevard which crowns the hill, has

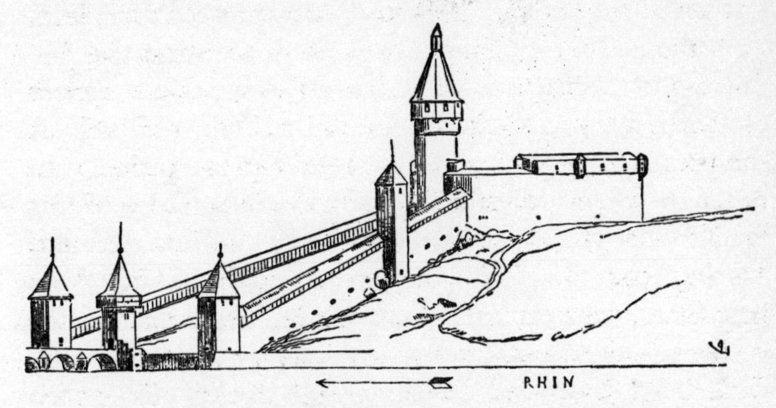

Fig. 99. Fortifications of the Bridge over the Rhine at Schaffhausen.

three heights of batteries, two covered and one open to the sky. The lowermost battery is placed a little above the bottom of the ditch, which is of great depth; we give the plan (fig. 100). The pentagonal *chemin-de-ronde,* A, is reached by a spiral ramp (or inclined way), B, of an easy incline, allowing the bringing up of cannon. At each of the angles of this chemin-de-ronde, which is about six feet six inches wide, are oblique embrasures for artillery commanding the ditch; in advance of the sides of the polygon are placed three small detached works, acting somewhat as bastions, of which (fig. 101) we give the perspective elevation. Supposing the besieging force to have suceeded in destroying one of those bastions by means of a breach battery placed on the counterscarp of the ditch (for the top of these bastions is no higher than the crest of the counterscarp, and they are completely masked from the outside), he would not have penetrated into the place; not only are those bastions detached, and with no communication except with the ditch, but they are armed with embrasures, C, for cannon at the gorges of the main work, pierced in the chemin-de-ronde (fig. 100), and their destruction would

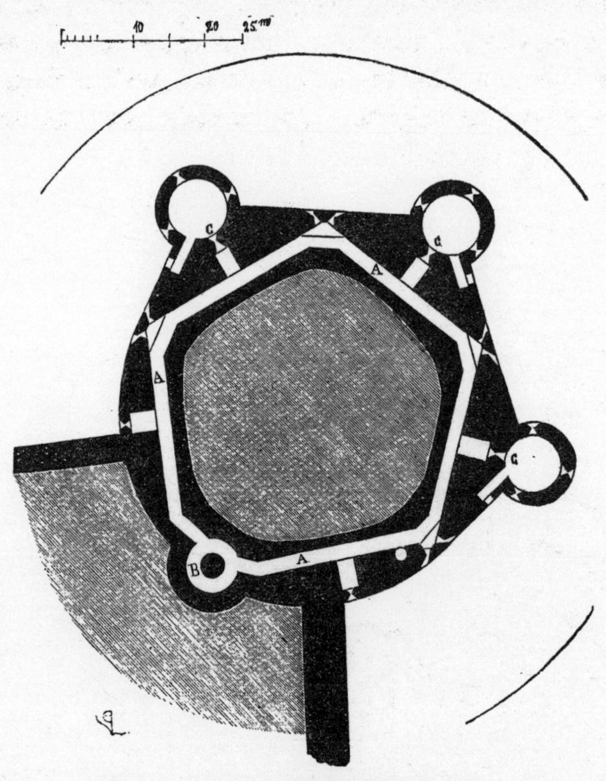

Fig. 100. Plan of the Citadel of Schaffhausen.

A A. Chemin-de-ronde. B B. Spiral Ascent. C C. Embrasures.

only serve to unmask these batteries. The bastions, built entirely of stone, are covered with cupolas having conical lanterns pierced with vent-holes to allow the smoke of the pieces to escape. The first story (fig. 102), which is reached by the same gentle spiral incline, B, in this case supported on four columns rising from the ground, shews, on the outside, a plan perfectly circular in form, the tower containing the inclined way being the only part projecting beyond the circle, on the side next the river. Towards the opposite point, at E, is a flying-bridge crossing the ditch; and on this side the architect has thought proper to strengthen his work by an enormous mass of solid masonry; not without reason, as it is only at this point that the fortress could be breached from the neighbouring heights. On the right of the boulevard,

Fig. 101. Perspective View of one of the Bastions, Schaffhausen.

higher up the river, at a point where an attack might also be attempted, is a casemated battery, F, separated from the principal hall by a thick wall of masonry. A breach made at G would not, therefore, admit the enemy within the works. H is an immense chamber, the pointed vaulting of which is supported on four massive cylindrical pillars. Four embrasures open out of this chamber,

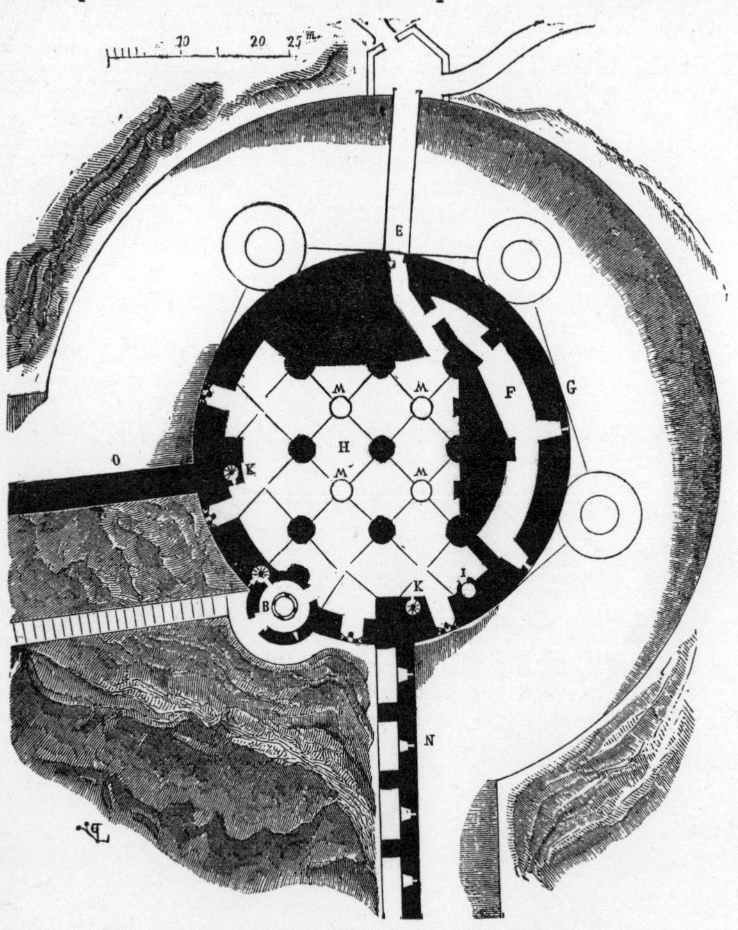

Fig. 102. Plan of the first Story of the Bastion, Schaffhausen.

B. Spiral ascent.
E. Flying Bridge.
F. Casemated Battery.
G. Outer Wall.
H. Vaulted Chamber, or Great Hall.
I. A Well.
K K. Small Spiral Staircases.
M M. Open Lunettes.
N O. Curtain Walls.

two flanking the two curtains which run down to the river, and two within the triangle. Besides the vent-holes pierced over each of the embrasures, four large lunettes, M, nearly three yards in diameter, are left open in the vaulting of the great chamber, for the purpose of affording light and air, and of allowing the smoke of the powder more rapidly to escape. At I is a well, and at K two small spiral stairs, communicating with the upper platform, for the use of the garrison. Close to the ramp is a third spiral stair ascending from the bottom. We give below (fig. 103) one of the embrasures of the great chamber, ingeniously planned to allow pieces of small

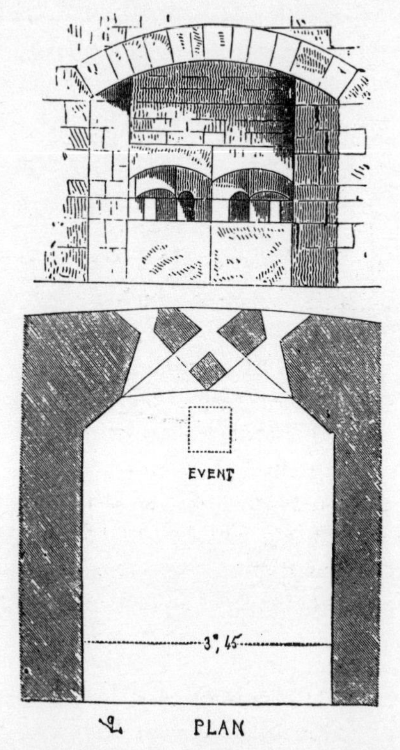

Fig. 103. Plan and Section of one of the Embrasures of the Great Hall.

calibre to fire in every direction without unmasking either those pieces or the men who served them. Fig. 104 is the plan of the upper story, or platform, the parapet of which is pierced with ten embrasures for cannon, and has four bartizans flanking the circumference of the fortress, having horizontal and descending loopholes, whereat to post arquebusiers. It will be seen that the two first embrasures to right and left command the interior of the triangle, and flank the tower which contains the ramp and which serves as a donjon and watch-tower for the whole work. The four lunettes, M, the well, I, and the small staircases for the use of the garrison, are repeated on this plan. The waters of the platform are carried off through ten gurgoyles placed under the embrasures. At N, O (fig. 102), are the two curtains which go down to the river. That marked N, up the river, is more strongly defended than the other; under the arches which carry the parapet and the wooden

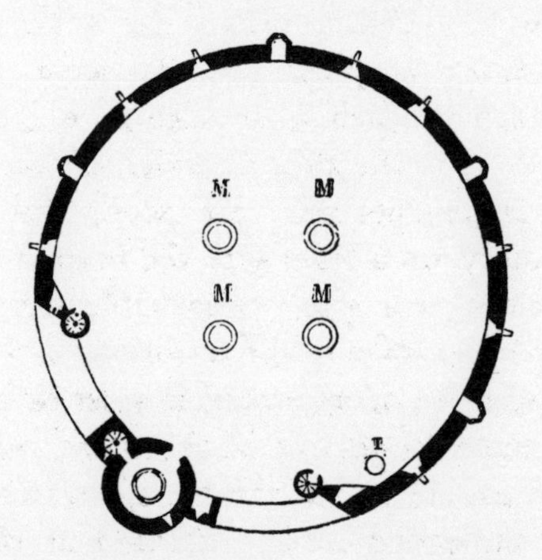

Fig. 104. Plan of the Platform of the Tower, Schaffhausen.

M M M M. Open Lunettes. I. The Well.

hoarding, still remaining in its original position, are pierced embrasures which command the slopes of the hill on the side where an enemy could present himself; the other side being defended by the walls of the faubourg of Schaffhausen. In order to give a clear notion of this fine fortress, as a whole, we give a view (fig. 105) taken within the triangle formed by the two curtains running down to the river. Here we see that the curtain, N, (the one which lies higher up the river) is flanked by a lofty square tower. We have restored the tower, which formerly stood at the head of the bridge, but is now destroyed. Of the works which surrounded this, there remain at present but a few traces. The ancient bridge has been replaced by a modern one. As regards the principal body of the fortress, the curtains, ditches, &c., nothing has been added to it, or taken from it, since the sixteenth century. The masonry is rude, but excellent, and has undergone no change. The vaults over the great hall are thick, well executed, and appear still to be bomb-proof.

This defence at Schaffhausen has a great aspect of power, nor have we preserved any work of the same period in France which is at once so complete and so ably planned. For the time at which it was erected, the flanking arrangements are very good, and the plan of the ground-story, on a level with the bottom of the ditch, is really set out in a quite remarkable manner. If we still find there the trace of the traditions of a fortification anterior to the use of ordnance, it must be allowed that the efforts made to get rid of these are very apparent, and the fortress of Schaffhausen appears to us to be superior to analogous works executed at the same period in Italy, which lays claim to having been the first to make use of the bastion.

It was not possible, however, to execute everywhere works of such importance or completeness. The object aimed at was rather the amelioration of the ancient defences, than their demolition to give place to new fortifications. In order to effect such sweeping changes, it

Fig. 105. Bird's-eye View of Schaff hausen.

would have been necessary for the engineers to have had at their disposal some fixed system, the goodness of which had been sanctioned by a long course of experience; but so far from that being the case, they proceeded only by a series of experiments, each engineer bringing forward his own observations and endeavouring to reduce them to practice. It is a striking fact that, after the wars of Italy, the French and Germans, having discovered that the Italian fortresses were narrow, circumscribed, and encumbered with defences which were rather in each other's way than of any mutual assistance, adopted in their new defences arrangements comparatively extensive in character, and endeavoured to fortify the outlying parts by boulevards of considerable diameter. In ordinary cases, and when the question was not so much to construct *de novo*, as to ameliorate fortifications already existing, while allowing the ancient system of defences to remain for the purpose of receiving bodies of archers, crossbow-men and arquebusiers, it was customary to erect *fausses braies* wherein batteries for horizontal fire could be planted, and which took the place of the lists which have been referred to in the course of this work. In pressing cases, the ancient walls and towers of the lists and the barbicans were simply taken down to the level of the parapet walk, and then crowned with embrasured parapets capable of receiving barbette batteries (fig. 106). The towers were looked upon so much in the light of an indispensable defence, and it was considered of so much importance to be able to command the open country, that they continued to be erected even after the use of *fausses braies*, arranged so as to flank the curtains, had been admitted. At first these *fausses braies* were given the same form, in plan, as had been given to the palisades, that is to say, they followed very closely the contour of the walls ; but they were soon converted into flanked works. The town

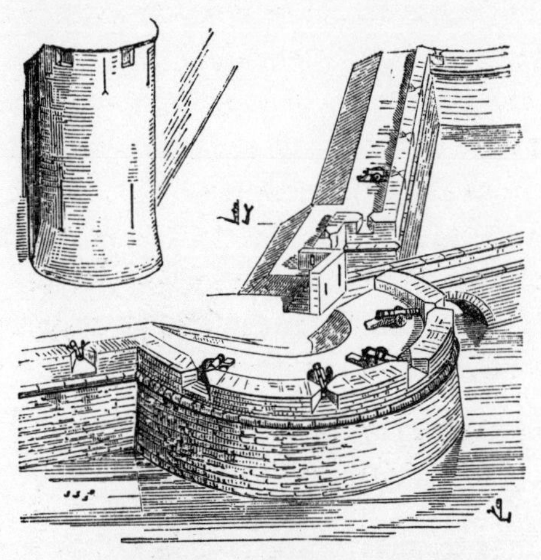

Fig. 106. View of a Battery, Schaffhausen.

of Orange was fortified anew under Louis XI., and already, at that period, the configuration of its defences was of this nature, (fig. 107). By means of these modifications, strong places were put into a state to resist artillery : this arm, however, was meanwhile undergoing rapid improvement. Louis XI. and Charles VIII. possessed a formidable artillery : the art of siege-warfare became every day more and more methodical ; engineers had adopted the system of regular approaches ; they had begun, when a fortified place could not be taken by sudden assault, to make trenches, to lay down parallels, and establish true siege-batteries, well gabioned. The old walls of the ancient defences being higher than the crests of the revetments of the ditches, offered an easy mark to the point-blank fire of the siege-batteries, and they could from a considerable distance destroy those uncovered works and effect a breach. In order to remedy this defect, the outsides of the ditches were furnished with palisades or parapets of masonry or timber, with earth-works and a first or external ditch; this

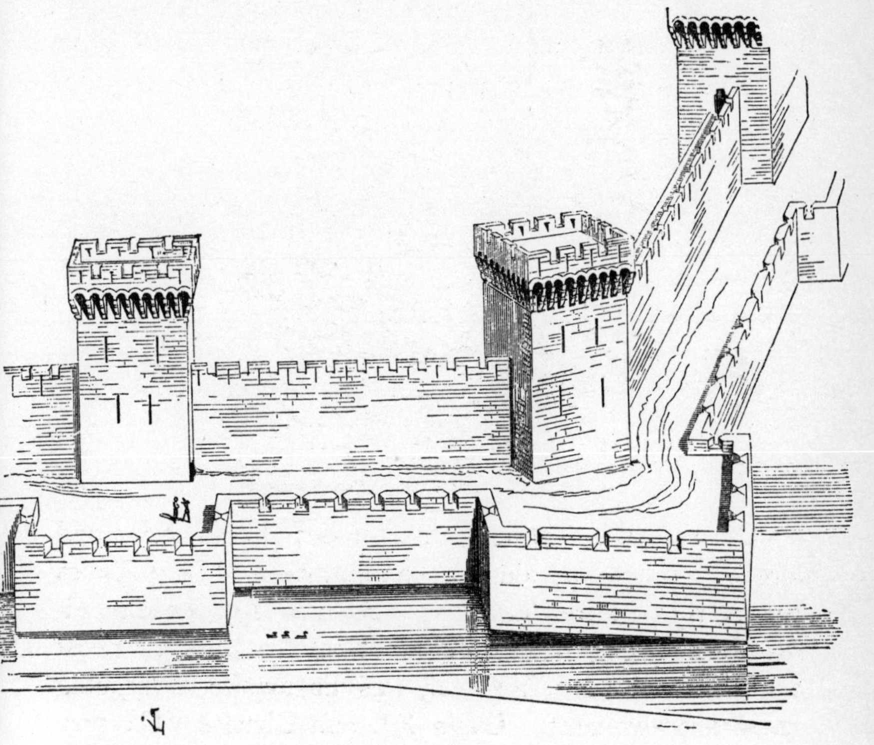

Fig. 107. View of part of the Fortifications of the Town of Orange.

work, which took the place of the ancient lists, preserved the name of *braie* (fig. 108). Outside the gates posterns and outworks were established, and earth-works, sustained by pieces of timber, were thrown up and were still named *boulevert bastille* or *bastide*. The description of the fortification of Nuys, which was besieged by Charles the Bold in 1474, explains perfectly the method employed for resisting attack[y] :—

[y] We borrow this passage from the "Historical Essay on the Influence of Fire-arms on the Art of War," by Prince Louis Napoleon Bonaparte, President of the Republic, p. 103. (Extr. from the Chronicle of Molinet, vol. v. ch. cclxxxiii. p. 42.)

"In like manner was Nuysse notably towered with free-stone, powerfully walled with works of exceeding strength, height, and thickness, and strengthened by strong *braiesses*, subtly constructed with stone and brick, and in divers places of

Fig. 108. Section of a Parapet at Orange, called a Braie.

earth wholly, adapted for defence by admirable artifices for repelling assailants, between which and the said walls there were certain ditches of no little depth, and in addition, before the said brayes, were other great ditches of an exceeding depth, whereof some were crested and filled with water to a great width, the which bathed the town and its forts as far as the river. Four principal gates of a like kind, together, and divers posterns and salients, embellished and fortified greatly the said enclosure; for every of them had in front its boulevert in the manner of a bastillon, great, strong, and defensible, furnished with all instruments of war, and sovereignly with great guns for powder [z]."

In this description we find the bastion standing clearly out as an important accessory of the defence, to fortify the salients, posterns and gates, and enfilade the ditches; and to take the place of the towers and barbicans of the lists in ancient fortifications, of the ancient detached

z "... traicts à poudre à planté."

bastilles, and of the defensive works outside the gates. In a little while this accessory, the importance of which was already seen, takes a principal place, and becomes at last the leading feature of modern fortification.

While they preserved, in the fortresses which were erected towards the close of the fifteenth century, the towers and curtains of the internal works, commanding a wide extent of country by their great elevation, which towers and curtains they further crowned with machicolations, they increased the thickness of the masonry, so as to render it capable of resisting siege-artillery, as we have shewn in the plans of the towers and boulevards of Langres and Schaffhausen. When the Constable of Saint Pol had the castle of Ham rebuilt in 1470, he not only thought it necessary to provide this fortress with advanced works and walls of counter-guard, but he had the towers and curtains, and more especially the great tower or donjon, built of such a thickness that those structures are still able to resist, during a considerable time, the force of modern artillery.

Up to the period at which we have arrived, it was in the details of the defence, the form and situation of the towers and curtains, that modifications had been introduced, with the view of meeting the new requirements; but the mode of construction in fortified works had undergone no change since the eleventh century. This consisted uniformly of two wall-faces of cut stone, brick, or rubble, enclosing a body or core of irregular rubble or stone concrete. Opposed to the sap or the ram, this sort of construction was good, for the pioneers found it more difficult to dislodge a mass of rubble, when the stone and mortar of which it was composed were hard and adhesive, than a structure built of regular masonry throughout; which, when a few stones came to be displaced, easily fell to pieces; regular masonry never preserving the homogeneity of well-built rubble-work. This kind of structure, therefore, resisted the shocks of the battering-ram more effectually than would a body of ashlar-work; but when pieces of ordnance came to replace the engines and instruments of destruction which had been employed in the middle ages, it was soon discovered that the stone revetments of these works, which were ordinarily from twelve to twenty inches in thickness, were readily shaken by the impact of iron balls; that they became detached from the internal core and left it exposed to the full effect of the fire, and that the stone merlons [a], when struck by cannon-balls, were shattered into fragments, and became themselves a means of destruction more deadly than the balls themselves. The architects of those works of defence, in order to prevent the shaking of the ancient walls and towers, strengthened the curtains by solid earth-works on the inside, and sometimes filled up solid the lower stories of towers. But when the wall fell under the fire of siege-artillery, those masses of earth tumbling foward with it, by forming a natural *talus* or incline, facilitated the access to the breach; while the simple walls, not backed by earth-works internally, when *they* fell, formed breaches of irregular configuration and of very difficult access. To remedy these defects, when the ancient defences were retained but adapted to act as a defence against artillery, the internal earth-works had occasionally wooden sleepers, or the branches of trees smoked or made resinous to preserve them from rotting, laid into the body of the clay in all directions; earth-works of this kind were firm enough

---

[a] The name given to those parts of the parapets between the crenelles or embrasures.

not to fall with the wall, and they rendered the breach impracticable. If it happened that the ancient walls had been simply backed up on the inside, so as to allow cannon to be placed on the parapet level, and the ancient crenellage had been simply replaced by thick merlons and embrasures of stone; then, when the besieged force had ascertained the point at which attack was threatened, and while the besiegers were making their last approaches and were engaged in effecting a breach, they threw up, in the rear of the front attacked, a work of timber and earth, and between this work and the breach they sank a ditch; the breach having become practicable, the besiegers threw forward their columns of attack, who now found themselves in front of a new rampart, well furnished with artillery; and the siege had to be begun afresh. This re-entering work was very difficult of access, being flanked by its natural conformation, nor could the assailants venture on carrying by assault a work which took them in face, in flank, and even in reverse. When Blaise de Montluc defends Sienna, he throws up behind the ancient walls of the city, and at the points where he supposes they will be attacked, re-entering ramparts in the manner of those sketched below (fig. 109).

"'So I had determined,' says he, 'that if the enemy came to assail us with artillery, to intrench myself at a distance from the wall, outside which the battery was placed, to let them enter at their ease; and I took care always to close the two extremities, and to place at each four or five pieces of heavy artillery, charged with thick chains, and nails, and pieces of iron. Behind this place of retreat (*retirade*) I decided on placing all the musketry of the town, together with the arquebusery, and when they should be within, to have the artillery and arquebusery to fire at once; and we, who would be at the two extremities, could then set upon them with pikes, halberds, swords, and bucklers [b].'"

Fig. 109. View of part of the Fortifications of Sienna.

This temporary defensive arrangement was not long in being established as a fixed system, as we shall presently see.

When the effects of artillery were well known, and it became an ascertained fact that walls of masonry of some

---

[b] *Comment.* of the Mareschal de Montluc, edit. Buchon, p. 142.

two or three yards in thickness (which was the mean thickness of curtains before the regular use of ordnance) could not resist a battery discharging from three to five hundred balls over a surface of eight yards square or thereabouts[c], at the same time that walls of masonry were lowered, various means were employed to give

---

[c] From the close of the sixteenth century, the French artillery had adopted six calibres for ordnance:—1st, the cannon, the length of which was 10 ft., with a ball weighing $33\frac{1}{3}$ lbs.; 2nd, the culverin, 11 ft. long, with a ball weighing $12\frac{1}{2}$ lbs.; 3rd, the *bâtarde*, $9\frac{1}{2}$ ft. long, with ball $7\frac{1}{2}$ lbs.; 4th, the *moyenne*, 8 ft. 2 in. long, with ball weighing $2\frac{3}{4}$ lbs.; 5th, the *faucon*, 7 ft. long, with a ball of $1\frac{1}{2}$ lbs.; 6th, the *fauconneau*, the length of which was 5 ft. 4 in., with a ball weighing 14 oz. (*La Fortification*, by Errard, of Bar-le-Duc, Paris, 1620.)

The following are some of the guns in use at the same period in England (see "A Military Dictionary explaining all Difficult Terms in Martial Discipline, Fortification, and Gunnery," by an officer who served several years abroad. London, 1702):—

Demi-cannon lowest. A great gun that carries a ball of 30 lbs. weight and 6 in. diameter. Its charge of powder 14 lbs. It shoots point-blank 156 paces. The weight of it is 5,400 lbs., the length 11 ft., the diameter of the bore $6\frac{2}{8}$ in.

Demi-cannon ordinary. A great gun $6\frac{4}{8}$ in. diameter in the bore, 12 ft. long, weighs 5,600 lbs., takes a charge of 17 lbs. 8 oz. of powder, carries a shot $6\frac{1}{8}$ in. diameter, and 32 lbs. weight, and shoots point-blank 162 paces.

Demi-cannon of the greatest size. A gun $6\frac{6}{8}$ in. diameter in the bore, 12 ft. long, 6,000 lbs. weight. The piece shoots point-blank 180 paces, 36 lbs. shot.

Demi-culverin of the lowest size: $4\frac{2}{8}$ in. diameter, 10 ft. long, 2,000 lbs. weight: shoots point-blank 174 paces, 9 lbs. shot.

Demi-culverin ordinary: $4\frac{1}{2}$ in. diameter in the bore, 10 ft. long, 2,700 lbs. weight: shoots point-blank 175 paces, 10 lbs. 11 oz. shot.

Demi-culverin, elder sort: $4\frac{2}{3}$ in. diameter, $10\frac{1}{3}$ ft. long, 3,000 lbs. weight: point-blank shot 178 paces, 12 lbs. 11 oz. shot.

There were also—

Culverin of the least size: bore 5 in. diameter, 4,000 lbs. weight; random shot, 180 paces, weight of shot 15 lbs.

Culverin ordinary: bore $5\frac{1}{4}$ in. diameter, 4,500 lbs. weight; carries a shot 17 lbs. 5 oz.

Culverin of the largest size: $5\frac{1}{2}$ in. diameter in the bore, 4,800 lbs. weight; carries a shot 20 lbs. weight.

Cannon Royal or of Eight: 8 in. diameter in the bore, 12 ft. long, 8,000 lbs. weight; weight of ball 48 lbs. Its point-blank shot is 185 paces.

There were other names given to pieces of ordnance, as:—Whole Cannon, Bastard Cannon or Cannon of Seven, Demi-cannon 24 pounders, Whole Culverin 12 pounders, Demi-culverin 6 pounders, Sakers, Minions 3 pounders, Drakes and Pedrerses.

It is worthy of remark that the range of point-blank fire then attained does not appear to have exceeded 190 paces.—TR.

them a greater force of resistance. In constructions ot a date anterior to the use of cannon, it had been sometimes customary, in order the better to resist the action of the mine, the sap, or the ram, to build in the thickness of the walls relieving or discharging-arches, masked by the outer face; which, by carrying the weight of the walls upon detached points, supported the parapets, and hindered the walls from falling all of a piece, unless it so happened that the besiegers had sapped them precisely at the concealed points of support (fig. 110), a casualty

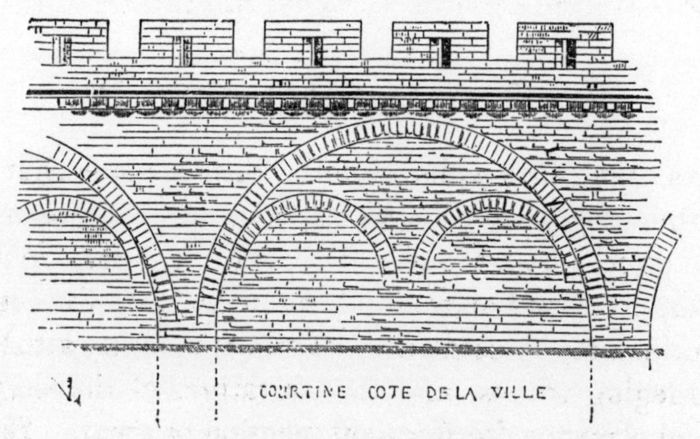

Fig. 110. View of the Parapet of the Curtain-wall, inside.

which could only be the effect of chance. In the sixteenth century this system was made perfect; for not only were discharging-arches built in the thickness of the curtains of masonry, but these were strengthened by internal abutments buried in the earth-works, and sustaining the revetments by means of vertical semicircular vaults (fig. 111). Care was taken not to connect these buttresses with the solid portion of the walls throughout their whole height, in order to hinder the revetments, when they fell by the action of the balls, from carrying the buttresses with them; these internal spurs could also, by sustaining the earth-work between them, offer an obstacle which it would be difficult to overthrow. But those means were costly; they always required,

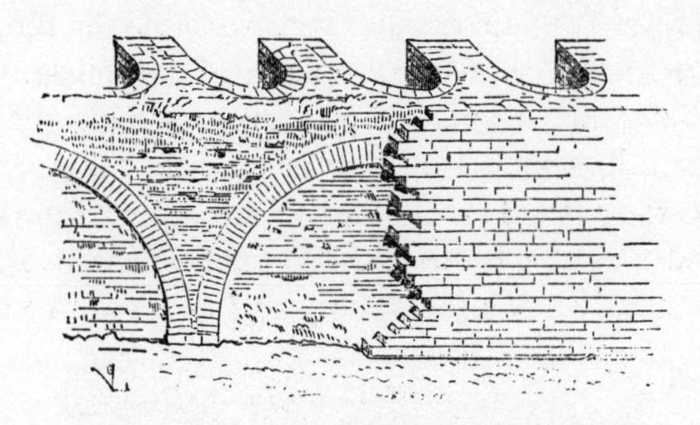

Fig. 111. View of a Parapet shewing the Construction.

besides, that the walls should form a somewhat considerable escarpment above the level of the counterscarp of the ditch. It was with difficulty engineers could be brought to abandon their elevated works; for, at this period, assault by escalade was still frequently attempted by besieging troops, and the narratives of the sieges of fortified places make frequent mention of them. Besides the means already described, whether for placing walls in a state to resist cannon, or for presenting a new obstacle to the besiegers when they had succeeded in overthrowing them, they did what was called *remparer* the fortifications, that is to say, they fixed on the outside of the ditches, or even as a protection to the wall to deaden the balls, or at a certain distance within the works, ramparts of wood and earth, the first forming a covered way, or a revetment to the wall, and the second a series of boulevards behind which to place artillery: 1stly, to embarrass the approaches and prevent a sudden assault, or to preserve the wall from the effect of cannon shot; 2ndly, to arrest the besiegers when the breach was effected. The first-named replaced the ancient lists, and the second obliged the besiegers to besiege the place anew, after the wall of enclosure had been destroyed. These ramparts of earth deadened the ball and resisted longer than walls of masonry; and they were better adapted to receive and to protect pieces in battery than the old earth-work parapets. They were constructed in several ways; the strongest were formed by means of an external revetment composed of vertical pieces of timber connected by St. Andrew's crosses, in order to hinder the work from undergoing displacement when some of its parts had been injured by the balls. Behind this timber-work facing was a series of *fascines* of small branches interlaced, or wattles, then an earth-work composed of alternate layers of wattles and earth. Sometimes the rampart was formed of two rows of strong stakes fixed vertically, bound together by means of flexible withes,

Fig. 112. Fascines.

and having an horizontal frame-work keyed in (fig. 112); the intervals being filled in with stiff clay well rammed down, with all the pebbles taken out, and interspersed with very small pieces of wood. Or else trunks of trees laid down horizontally, connected together by cross-pieces keyed through, and with the intervals filled in as last described, formed the rampart (fig. 113). Embrasures were left at intervals, with hanging flaps. If the besieged were attacked suddenly, or if they could not obtain the kind of clay required, they contented themselves with binding together trees which retained a portion of their branches, the interspaces being filled in with fascines (fig. 114[d]). Those new impediments opposed to siege-artillery led to the use of hollow balls and pro-

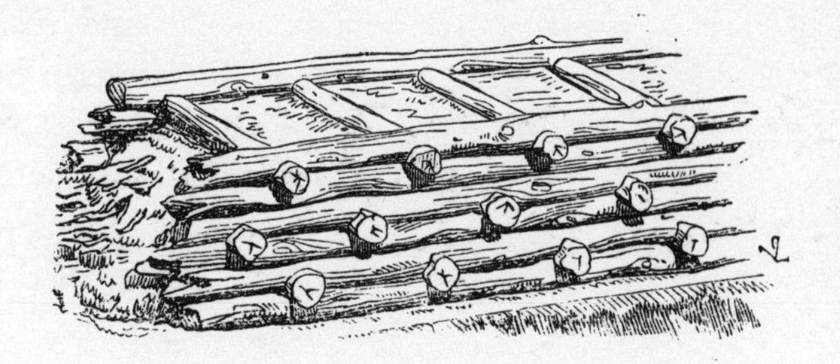

Fig. 113. Rampart formed of the Trunks of Trees.

jectiles charged with combustibles, which, exploding in the midst of the ramparts, produced great disorder. By

Fig. 114. Rampart formed of Branches of Trees.

degrees, sudden overt attacks had to be abandoned, and places thus guarded approached only under cover, and along winding trenches, the angular or rounded turnings of which were protected from enfilade fire by gabions filled with earth and set on end. These large gabions served also for masking pieces placed in battery; the

---

[d] See *Le roi sage, Récit des Actions de l'Empereur Maximilien Ier*, by Mark Treitzsaurwen, with the engravings of Hannsen Burgmair, published in 1775; Vienna. (The engravings in wood of this work date from the commencement of the sixteenth century.)

intervals between the gabions forming the embrasures (fig. 115 [e]). When the besiegers, by means of trenches,

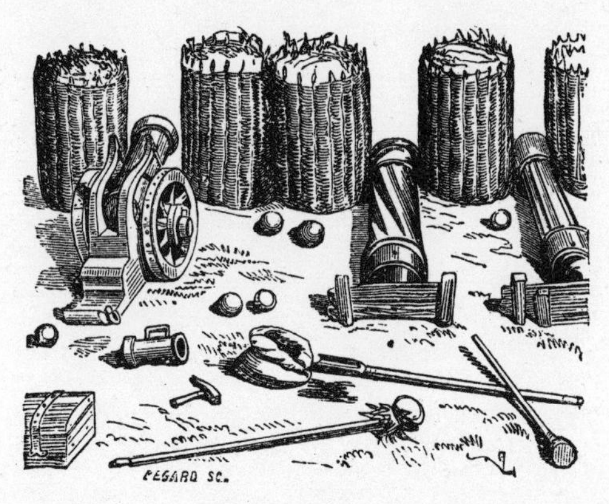

Fig. 115. Embrasures formed with Gabions.

succeeded in placing their last batteries close up to the fortifications, and these latter were furnished with good external ramparts and with walls of great elevation, it became a matter of necessity to protect the breach battery against the horizontal and plunging fire of these works, by embankments of earth surmounted with rows of gabions or of palisades strongly bound together and lined with wattles. Those works could only be executed during the night, as it is the practice still to execute them (fig. 116 [f]).

Whilst thus improving the defensive works, by strengthening the walls with ramparts of timber or earth on the outside of the ditches, or against the outer face of the walls themselves, it was felt that these means, although rendering the effect of the fire of the artillery less terrible and immediate, still could do no more than retard the assault by a few days; and that a fortified

---

[e] See the note on the preceding page. [f] Ibid.

place, when once invested, and with breach-batteries fixed within a short distance of its walls, found itself

Fig. 116. View of the Trenches, with Gabions, &c.

shut up within these walls, without being able to attempt sorties or to hold any communication beyond. In conformity with the method hitherto in use, the assailants still, at the close of the fifteenth century and the beginning of the sixteenth, directed all their efforts against the gates; the ancient barbicans, whether of stone or of wood (*boulevards*), were no longer either spacious enough or sufficiently well flanked, to oblige the besieging force to undertake any great works of approach; and they were easily destroyed: whilst, once having effected a lodgment in these outworks, the enemy

Fig. 117. The Mazelle Gate and Barbican at Metz.

A. Barbican. B. Ancient Curtain. C. Later Curtain.

fortified himself in his position, established his batteries, and concentrated his fire upon the gates. These were, therefore, the first points upon which the attention of constructors of fortifications became fixed. From the close of the fifteenth century the great object aimed at was to guard the gates and the *têtes-de-pont;* to flank those gates by defences adapted to receive artillery, taking as much advantage as possible of the existing defences and improving them. The Mazelle gate (fig. 117) of the city of Metz[g] had been strengthened in this manner; the ancient barbican, at A, had been levelled and terraced to take cannon; the curtain-wall, B, had been rampired (*rampard*) on the inner side, and the one, C, reconstructed so as to command the first gate. But defences so narrow and restricted did not suffice; those who conducted the defence were in each other's way; while the siege-batteries, when brought into position before works thus accumulated on a single point, destroyed them all at the same moment, and flung the defenders into disorder. Engineers soon obeyed the necessity there was for widening the area of defence, and extending the works so as to command a greater space of ground. To this end, they erected boulevards outside the gates, in order to afford the latter shelter from the effects of artillery (fig. 118[h]); sometimes these boulevards were furnished with *fausses braies* to receive arquebusiers; if the enemy, after having destroyed the merlons of the boulevards and dismounted the batteries, reached the ditch, the arquebusiers retarded the assault. A great extension, also, was given to the external works in order to form *places d'armes* in front of the gates. The increasing power of artillery led, as its result, to the gradual extension of the fronts of fortifications, and to

---

[g] Mazelle Gate at Metz. (Topog. of Gaul, Mérian, 1655.)
[h] Gate of Lectoure. Ibid.

their passing beyond the limits of the ancient walls and towers, to which tradition, quite as much as any motive

Fig. 118. View of Barbican, or Boulevard, Metz.

of economy, had at first confined them. The towns were attached to their old walls, and could not be induced all of a sudden to look upon them as defences all but valueless; if necessity required they should be altered, this was almost always effected by means of works of a provisional character. The new art of fortification was still but in its elements, and each engineer endeavoured, by experiment, not indeed to establish a system which should be original and universal, but to preserve the ancient walls of his town by intrenchments, partaking more of the character of field-works than of that of a system of fixed defences, methodically planned. These various experiments, however, would necessarily lead to a general result: the ditches were in a short time carried round the boulevards of the gates, front and rear, in a manner which had already been adopted at some barbicans; and on the outside of these ditches, ramparts of earth were thrown up forming a covert-way. It was thus by slow degrees that engineers succeeded in commanding the approaches of the besieging force. The want was felt of fortifying the outlying parts, of protecting towns by works of sufficient projection to hinder siege-batteries from bombarding the dwellings and stores of the besieged; and it was more especially along navigable rivers and sea-ports that they had already in the fifteenth century begun to plant towers (or bastilles) connected by ramparts, in order to place the ships out of the reach of projectiles. The towns of Hull in Lincolnshire, of Lubeck in Holstein, of Leghorn, of Bordeaux, of Douai, of Lièges, of Arras, of Basle, &c., possessed bastilles capable of receiving cannon. We subjoin the plan of the line of towers of Kingston-upon-Hull, reproduced by Mr. J. H. Parker (fig. 119[i]). As regards the bastilles of Lubeck, they were detached, or connected with *terra firma* by jetties, and thus formed salients of very considerable projection surrounded on all sides by water (fig. 120[k]). These latter bastilles appear to have been constructed of timber and earth.

---

[i] Some Account of Domest. Architect. in England, from Edward I. to Richard II.; Oxford, J. H. Parker, 1853. The castle of Kingston-upon-Hull was founded by King Edward I. after the battle of Dunbar, but the fortifications here figured are certainly of a later date, and belong probably to the close of the fifteenth century. Mr. Parker remarks with justice that they were in conformity with the external defences adopted in France.

[k] After an engraving of the sixteenth century from the author's collection.

The method of defending gates by bastions, or boulevards, of a circular form, was employed in France from

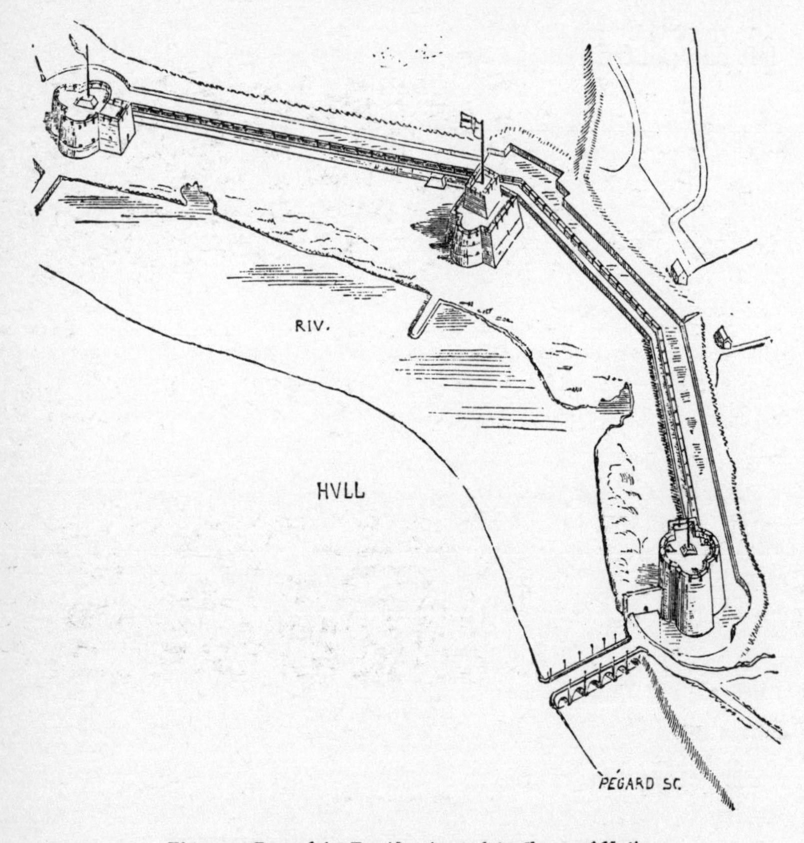

Fig. 119. Part of the Fortifications of the Town of Hull.

the time of Charles VIII. Machiavelli, in his "Treatise on the Art of War," l. viii., thus expresses himself:—

"But . . . if we have anything considerable (in the way of military institutions) we owe it entirely to the ultramontanes. You know, and your friends can remember, what was *the state of weakness of our fortifications* before the invasion of Charles VIII. in Italy, in the year 1494."

And in the official account of his visit of inspection

to the fortifications of Florence, in 1526, occurs the following passage :—

"We arrived afterwards at the gate of San-Giorgio (on the left bank of the Arno) ; the advice of the captain was to lower

Fig. 120. Fortifications of Lubeck.

it, to form it into a round bastion and to place the outlet on the flank, as is the custom."

We give (fig. 121) a bird's-eye view of the castle of Milan, as it was at the beginning of the sixteenth century, which will serve to explain the system of defence and attack of fortified places in the time of Francis I.

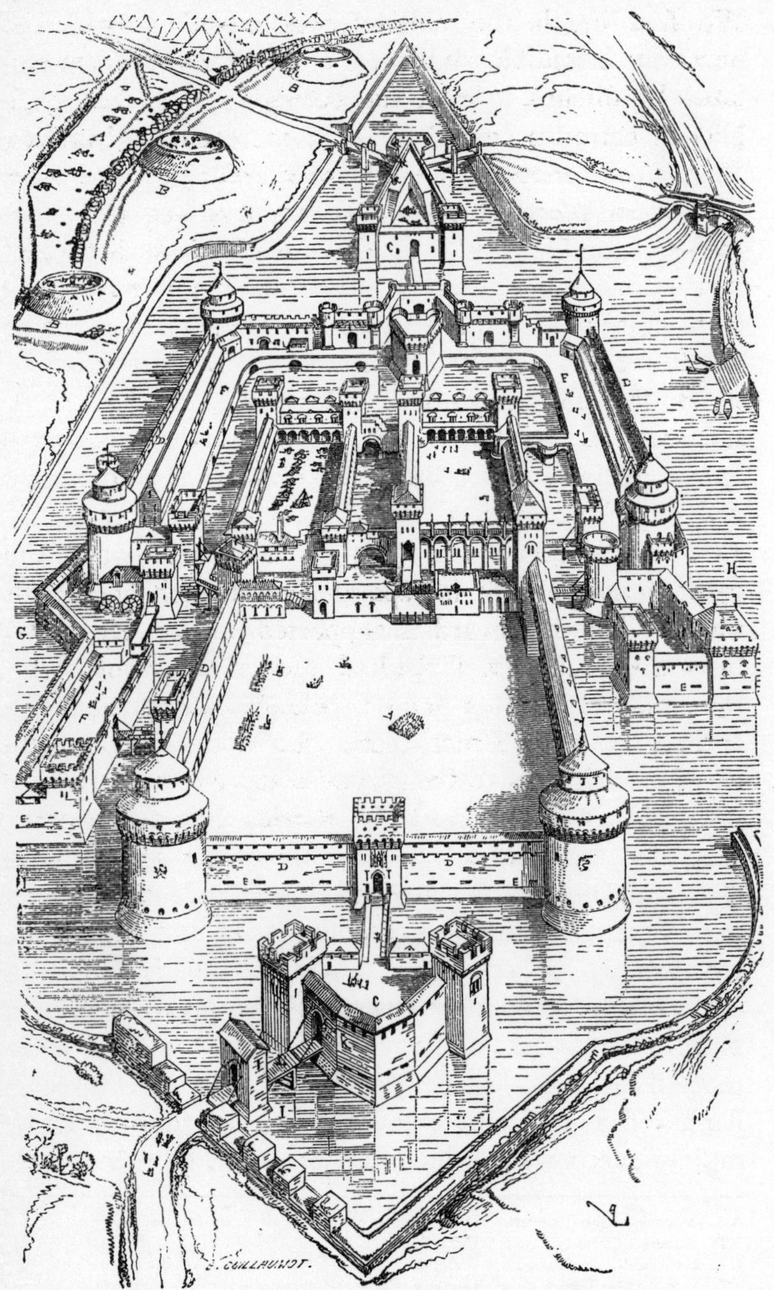

Fig. 121. Bird's-eye View of the Castle of Milan.—(See opposite.)

We find herein the old defences mixed up with the new, an incredible confusion of towers and forts isolated by ditches. At A the besieging army had established batteries behind gabions, protected by bastilles, B, circular redoubts of earth-work occupying the place of modern *places d'armes*, but commanding the front works of the besieged. At C we see boulevards flanked by towers in advance of the gates; at D curtains without terraced-work, but crowned with parapets (*chemins de ronde*); on the ground-floor are placed covered batteries, the embrasures of which are seen throughout, at the level, E, whilst the upper portions of the walls appear entirely devoted to the archers, crossbow-men, and arquebusiers, and still retain their machicolations. At F is a boulevard surrounding the weakest part of the castle, from which it is separated by a ditch filled with water. This boulevard is supported on the left, at G, by a work rather well flanked, and on the right, at H, by a kind of keep or donjon, defended according to the ancient system. From these two works the communications with the body of the place are by means of drawbridges. The castle is divided into three parts, separated by ditches, and capable of cutting off their communications with each other. In advance of the gate situate in the foreground, at I, and along the counterscarp of the ditch, a parapet wall is run, with traverses, to prevent the besiegers from taking the flank, K, *en écharpe*, and destroying it. But it is easy to see that all these works are too small, and do not present flanks of a sufficient extent; and that they could be rapidly overpowered, one after the other, if the besiegers

A A. Batteries of the Besiegers.
B B. Bastilles of ditto.
C C. Boulevards, or Barbicans, flanked by Towers.
D D. Curtain Walls.
E E. Embrasures.
F F. Boulevards.
G. Outworks.
H. Donjon, or Keep.
I I. Parapets.
K. The Flank.

were in possession of a numerous artillery, the force of which might be converged upon one point after another, simply by changing the direction of the fire. And, in fact, in order to prevent works like these, of too limited an extent, from being all destroyed at the same time by a single battery which could be brought to bear on them from a point sufficiently near, they had already begun to erect, within the fortifications and in the midst of the bastions, earth-works of a square or circular form, for the purpose of commanding the earth bastilles of the besieging force. This kind of work was frequently employed in the sixteenth century, and since, and took the name of *cavalier*, or platform; it became a resource of great utility in the defence of strong places, whether it was of a permanent nature, or was merely erected during the course of the siege; enabling the besieged to sweep the trenches, to take the siege-batteries *en écharpe*, or to command a deep breach when the embrasures on the flanks of the bastions had been destroyed by the enemy's fire. As permanent works, platforms were frequently erected for the purpose of commanding roads, gates, and especially bridges, when these latter, on the side opposite the town, opened upon the bottom of an escarpment whereon the enemy could establish batteries to protect an attack, or to hinder the besieged from establishing themselves in force at the other side. The bridge at Marseilles, spanning the ravine which formerly intersected the Aix road, was defended and enfiladed by a great cavalier, or platform, situate on the town-side (fig. 122[k]). When the bastions were too distant one from the other to flank the curtains effectually, platforms were erected between them in the centre of the curtains

---

[k] *Vue de la ville de Marseille.* (Topog. de la Gaule, Mérian.)

either semicircular or square in form, to strengthen their fronts; and in connection with the bastions themselves

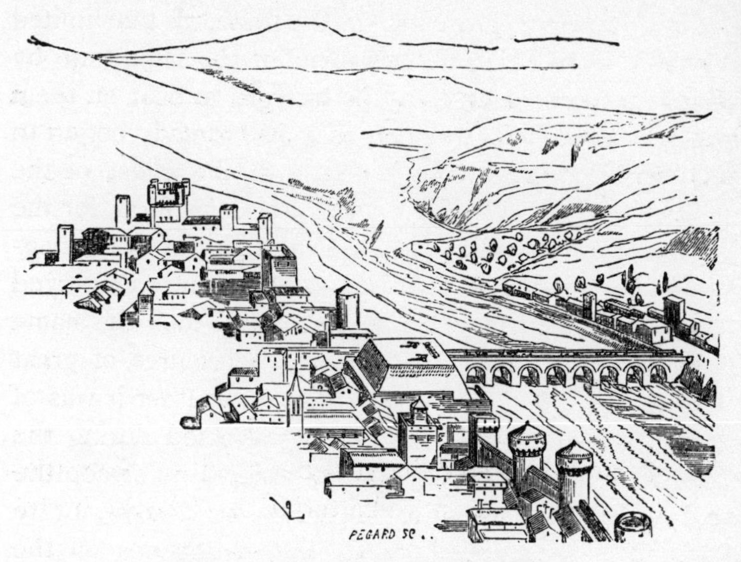

Fig. 122. View of the Bridge of Marseilles.

it was usual to erect platforms, in order to give the bastions a greater commanding force, and to enable a greater number of pieces to be placed in battery on a given point. In the white marble bas-reliefs which decorate the tomb of Maximilian at Innspruck, we see a cavalier, or raised platform, planted on a bastion forming part of the fortifications of Verona (fig. 123). The bastion is well characterized, with its faces and flanks; a *fausse braie* defends the lower portions of it and commands the ditch. The parapets are faced with earth and branches; behind are gabions to protect the soldiers; above the gabions, on the terre-plain of the bastion, rises a platform, or cavalier, built in masonry, the parapets of which are, in like manner, furnished with fascines and earth.

Platforms had this further advantage, that they defiled the curtains; which was all the more necessary as

Fig. 123. Cavalier on a Bastion at Verona.

besieging armies still preserved, at the beginning of the fifteenth century, the traditions of the offensive bastilles of the Middle Ages, and frequently established their siege-batteries upon earth-works raised considerably above the surrounding ground. When the besieging force, either by means of earth-works, or owing to the configuration of the ground outside, were able to plant their batteries upon some elevated point commanding (or level with) the upper defensive works of the fortifications, and, taking them *en écharpe* or enfilading them, could thus destroy the uncovered batteries of the besieged, at a long range and over a wide extent, it was usual from the date of the sixteenth century to erect, in default of platforms, traverses of earth, A (fig. 124), sometimes

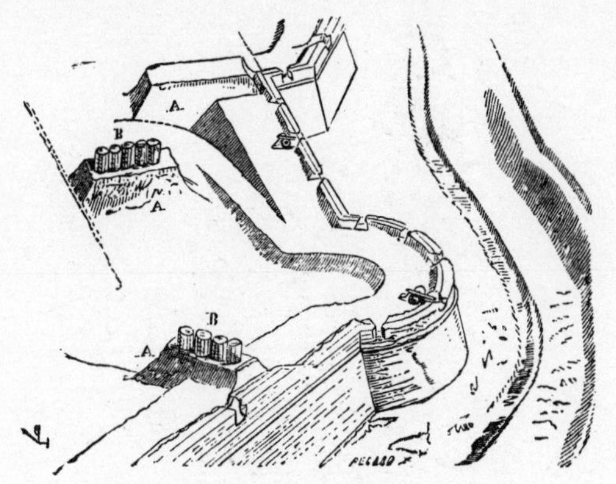

Fig. 124. Traverses, A, with Gabions, B.

furnished with gabions, B, at the moment of the attack to increase their height.

But it was not long before the defective nature of works, which, while they formed considerable salients from the outer lines, did not connect themselves with any general system of defence, came to be recognised: they were not flanked. Obliged to defend themselves, separately, and not being themselves defended, they merely presented a single point upon which the fire of the besiegers could be converged, and could only oppose an almost passive resistance to the cross-fires of the siege-batteries. By increasing his obstacles, they retarded the works of the enemy, but were impotent to destroy them; the bastions or platforms, therefore, were multiplied; that is to say, that instead of raising them only in advance of gates, or, as at Hull, with a special purpose, they erected them at regular intervals, not only to keep the approaches at a distance and cover the ancient fortified fronts, still preserved, from the fire of the enemy, but also in order to defend these bastions by means of one another[1]. In the *procès-verbal* drawn up by Machiavelli, and already referred to, on the fortifications of Florence, we find the following passages, having reference to the erection of round bastions in advance of the ancient fortified fronts :—

"When you go beyond the road of San-Giorgio about one hundred and fifty *braccia* (or about one hundred and thirty yards), you come upon a re-entering angle, formed by the wall which at this point alters its direction and turns to the right. The opinion of the captain (general) was, that it would be useful to erect at this point either a casemate or a round bastion which should command the two flanks; and you will understand that he means by this, to sink ditches wherever there are walls, because he is of opinion that ditches are *the first and the strongest defences* of fortified places. After having advanced a distance of some one hundred and fifty *braccia* further, to a place where there are some buttresses, he was of opinion there should be another bastion erected here; and he thought that if this were made of sufficient strength, and sufficiently advanced, it might render unnecessary the erection of the bastion of the re-entering angle, already referred to.

"Beyond this point we find a tower, whereof he considers we should increase the extent and diminish the height, arranging it in such a manner as to be able to work heavy guns upon its summit; he thinks it would be useful to do the same with all the other towers; he adds that the nearer they are to one another the more they increase the strength of a place, not so much because they reach the enemy in flank as because they attack him in front."

In nearly all cases these boulevards, or bastions (for we may henceforward give them that name), erected

[1] *Défenses de la ville de Blaye*, (Topog. de la Gaule, Mérian). The plans of the towns of Tréves, Dôle, Saint-Omer, Douay, &c., (see *Les Plans et Profils des Princip. Villes*, by the Sieur de Beaulieu, 17th century). The plans of the towns of Bordeaux, Mons, Liége, Coblentz, Bonn, Basle, (see Introd. to *La Fortification* of De Fer, *Atlas ïtal.*, 1722).

precipitately and on the spur of the moment, during a siege, outside the ancient fronts, were merely earthworks faced with timber or sods, and not rising above the crest of the counterscarp of the ditch. But when, during the first half of the sixteenth century, the ancient towers and curtains of masonry were replaced by new defences, and when it happened that those new works could be carried out (the necessary funds being forthcoming) methodically, those works were then revetted with masonry. Up to this period, however, they did not attempt to extend the works outside the body of the place, and the attack could always establish itself opposite the bastions of the fortifications, without being obliged to take a certain number of outworks, such as those it is now customary to dispose around the main walls. In order to oblige the besiegers to commence their approaches at a sufficient distance from the glacis, recourse was had to platforms (or *cavaliers*) high enough to command the country, or else to towers so arranged as to afford a view of the neighbourhood of the fortifications over the top of the curtains and bastions. It was according to this method that Albert Durer fortified the town of Nuremberg. Whilst retaining the ancient defences of the town, which date from the fourteenth and fifteenth centuries, Albert Durer erected outside these an advanced line of earth-work fortifications, with large revetted circular bastions, ditches, and glacis, and at intervals he strengthened the defences of the old walls by lofty and spacious towers, commanding the two lines of fortification and the outlying parts. Those large towers are furnished on the ground-story with one or two embrasures to enfilade the parapet between the ancient and the new lines, and have a platform at top capable of receiving a considerable number of cannon.

No fortification of this period shews more forcibly what occupied the attention of the engineers of the sixteenth century than those Nuremberg works. Not having as yet adopted a complete system of flank-works; not having as yet applied, with all its consequences, the axiom that *that which defends should be defended;* uneasy about the result of their combinations, or rather, we should say, of their experiments; knowing that the convergence of an enemy's fire would always destroy their advanced works, however solidly they might be built,—they wanted to see and command their external defences from within the body of the place, as, a century before, the baron could see from the top of his donjon whatever was going on round the walls of his castle, and send up his supports to any point attacked. The great towers of Nuremberg are more properly, in fact, detached keeps than portions of a combined system; rather observatories than effective defences, even for the period. The plunging fire from those platforms, placed at a height of some eighty feet above the level of the parapets of the town walls, could produce no very great effect, more especially if the besieging force succeeded in establishing itself on the crest of the counterscarp of the ditches, for from that point the swell of the bastions masked the towers. It does not appear that those lofty towers were very frequently employed. The attention of the engineers of the commencement of the sixteenth century was principally fixed upon the form to be given to the newly adopted bastions, and to the principles to be adopted in their construction and defence. While giving them a height sufficient to command the surrounding parts, and a great diameter as compared with that of the ancient towers attached to outer embanked walls; while strengthening the masonry of them by buttresses which were covered by the embankments behind; and while projecting them before the curtains as much as was possible without detaching them altogether,—they endeavoured, at first, to protect their anterior part from the converging fire of an enemy's batteries. For this purpose they threw up around the circular bastions, and at their base, *fausses braies*, masked by the counterscarp of the ditch; and these, to render them stronger, were sometimes flanked. This already shewed a marked progress, for circular bastions, like circular towers, were weak if attacked on the face; they could oppose to the converging fire of a breach battery only one or two pieces of cannon. We give below an example of one of those flanked *fausses braies* (fig. 125[m]).

---

[m] *Della Cosmog. universale*, Sebast. Munstero, 1558, small fol. *La citta d'Augusta* (Augsburg), p. 676. The bastion here given is a dependency of a very important advanced work, which protected an ancient front (of the old walls) built behind a wide wet ditch. The curtain, G, is feebly flanked by the bastion, because it is commanded and enfiladed in its whole length by the old walls of the town; as regards the curtain, H, it was flanked by the fausse braie, and by the bastion, E, prolonged. It was with difficulty the bastion could be attacked behind the flanks of the fausse braie, at D, and it was impossible to attack it on the side of the curtain, G, for then the besiegers found themselves taken in reverse by the artillery posted on the old ramparts which commanded the flank, I, of the bastion. Engineers had already begun to apply with some degree of method the principle, that *the parts inside should command the parts outside*, and the assailants having obtained possession of the bastion, found themselves exposed to the fire of a very extended line of front (see fig. 126). A is the front of the old walls, altered for cannon; B, a wide watercourse; C, a covered way with barrier, embanked against the advanced work; D, a small watercourse; E, traverses; F, bridges; G, a rampart crossing the ditch, but commanded, enfiladed, and taken in reverse by the old walls, A, of the town; H, the advanced work; I, a front of old wall, levelled and rampired; K, a front also rampired (these two low ramparts are commanded on all sides by the walls of the town); L L are bridges, M is the wet ditch, N the bastions in earth, timber, and wattles, one of which is detailed in fig. 125; O the remains of the old embanked defences; P the covered ways of the advanced work. (See the general plan of the town of Augsburg, which shews a series of bastions constructed according to this system (fig. 127).—Introd. to *La Fortification*, dédiée à Monsiegneur le duc de Bourgogne, Paris, 1722, fol. Ital.) This plan, which is more accurate than that given in the work of Séb. Munstero, differs, as regards the advanced work, H, in some important particulars. We think it worth while to give them both, especially as the principle of defence obtained by the

When the besiegers had destroyed the battery established at A and completed their approaches, having

Fig. 125. View of one of the Bastions of Augsburg.

A. The Battery.
B. The Glacis.
C C. Covered Way.
D. Fausse Braie.
E. Prolongation of Bastion.
G H. Curtains.
I. Flank of the Bastion.

gained the top of the glacis at B, they had to drive back the defenders of the covered way, who were protected by an embankment and a palisade; if they succeeded in

addition of the bastions of the sixteenth century is preserved in both plans. We can still trace at Augsburg the lines of those bastions, the fausses braies of which, in masonry, have been enclosed within an external sodded embankment.

reaching the ditch, they were received by the horizontal and cross-fire of two pieces placed in the flanks of this

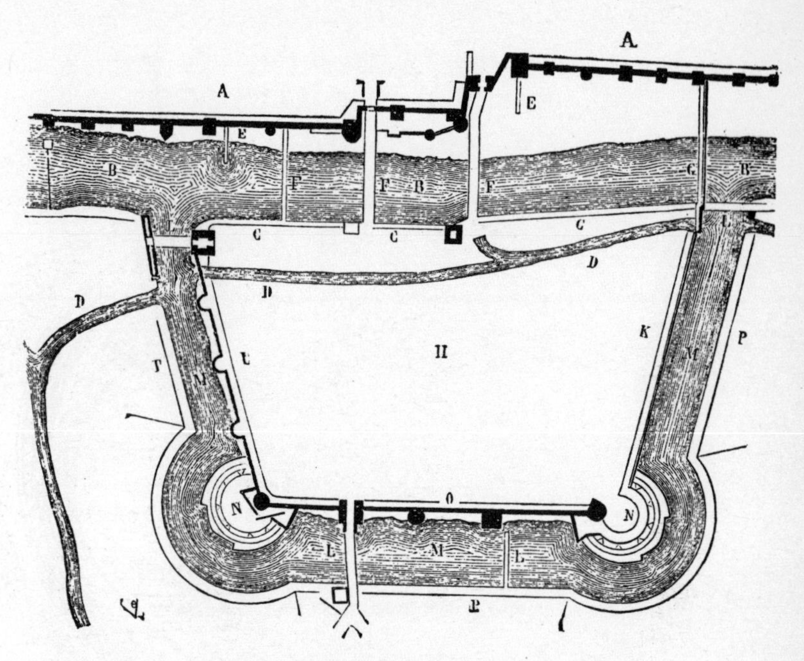

Fig. 126. Plan of Bastions at Augsburg.

A. Front of old Wall.
B. Wide Watercourse.
C C. Covered Way.
D. Small Watercourse.
E. Traverse.
F F. Bridges.
G. Rampart across Ditch.
H. Advanced Work.
I K. Fronts of old Wall with Ramparts.
L L. Bridges.
M. The wet Ditch.
N N. Bastions.
O. Remains of old Defences.
P. Covered Way of the Advanced Work.

lower work (which was preserved until the moment of attack by the fausse braie at C), and by the musketry of the defenders of the counterscarp of the ditch. To fill the ditch under the cross-fire of those two pieces was an operation attended with great danger; it then became necessary to destroy the fausse braie and its flanks by cannon. If the assailants wanted to turn the flanks and take the fausse braie at D by escalade, they were received by the masked pieces of the second flank, E. Finally, having overcome all those obstacles and carried the bastion, the assailants found still before them the old defences, F, which had been preserved and raised upon; the lower portions of which, masked by the elevation of the bastion, could be furnished with artillery or arquebusiers. The artillery, also, intended to command the curtains when

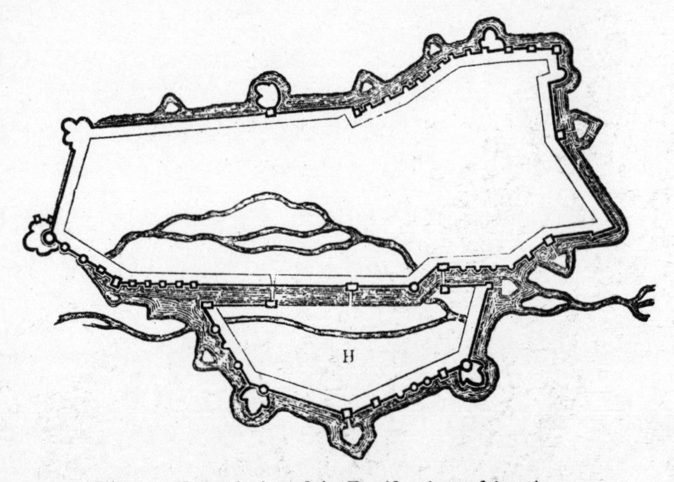

Fig. 127. Ground-plan of the Fortifications of Augsburg.

these latter were destroyed and the assailants attempted the passage of the ditch to mount the breach, were masked. In order to obtain this result, the engineers of the sixteenth century gave, as we have already explained, a great projection to their round bastions from the face of the curtains, in such a manner as to form a re-entering angle which was pierced with embrasures for cannon (fig. 128[n]). But more space was required in

[n] *Della Cosmog. universale*, Sebast. Munstero, 1558, small folio. *Sito e fig. di Francofordia città, come e nell' anno* 1546. The bastion drawn in this view commands the river (the Maine) and one entire front of the ramparts of the city. This fortified angle is very interesting as a study, and the engraving which we have

Fig. 128. A. The Re-entering Angle, with a Casemated Battery. B. Masking Wall. C C. Vent-holes. D. The Rampart.

Fig. 128. View of part of the Fortifications of Frankfort-on-the-Maine.

the gorges, A (fig. 129), for the service of the artillery; their narrowness rendered them difficult of defence when the enemy, after having seized upon the bastion, attempted to push forward. We have seen how difficult it was, before the invention of cannon, to oppose to an assaulting column, narrow but deep, precipitated upon the parapets, a defensive front sufficiently solid to drive the assailants back (fig. 21); and when artillery opened large practicable breaches in the bastions or curtains, owing to the falling in of the earthworks, the assaulting columns could thenceforward be not only deep, but also present a large front: it thus became necessary to oppose

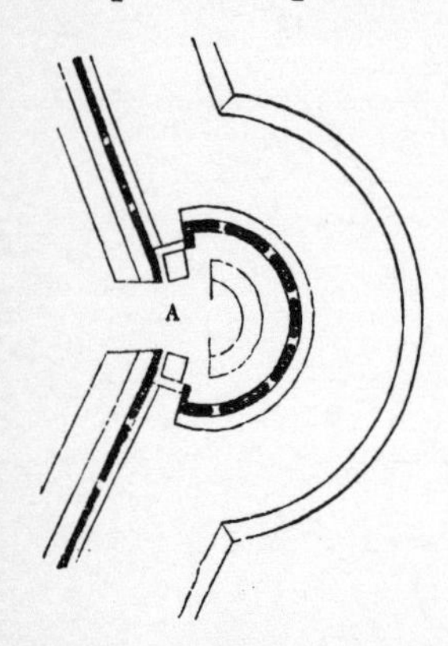

Fig. 129. Plan of one of the Bastions.

copied, with such emendations as were required to render it clear, indicates the various modifications and improvements which had been introduced into the defence of strong places at the beginning of the sixteenth century. In the centre of the new bastion the ancient angle tower (*tour du coin*) has been preserved, and serves as a watch-tower; this tower has evidently had the top story added in the sixteenth century. The bastion is armed with two heights of batteries, the lower of which is covered and masked by the counterscarp of the ditch, made in the manner of a wall of counterguard. This covered battery could not be used until the moment when the assailants had attained the ditch. The re-entering angle, A, which contains a casemated battery, is protected by the projection of the bastion and by a wall, B, and commands the river. Vent-holes, C, allow the smoke from the batteries to escape. Beyond the drawbridge is a rampart erected in advance of the old walls, and commanded by them and the towers; it is guarded by a fausse-braie intended to protect the passage of the ditch. Arched buttresses are visible at intervals, which abut at one side against the revetted wall of the rampart and slope to the fausse braie; this latter is enfiladed by the fire of the angle-bastion, and by a re-entering angle of the rampart, D. Were it not for its narrow limits, this defence might still be considered as of considerable strength. We have thought it right to admit various examples which do not belong to the military architecture of France; for it must be admitted that at the time of the transition from the ancient to the modern system of fortification, the several Western nations of Europe rapidly adopted the new improvements introduced into the art of defending strong places, and that local traditions were forgotten when necessity had become the teacher.

to them a front of defenders at least equally great, to prevent the latter from being outflanked. The narrow gorges of the ancient circular bastions, although well closed with ramparts internally, were easily carried by assaulting columns, the impulsive force of which is always very powerful. The grave defects attendant upon narrow gorges was soon perceived, and in place of retaining the circular form in bastions, they were given (fig. 130) a straight face, B, and two cylindrical ends, C,

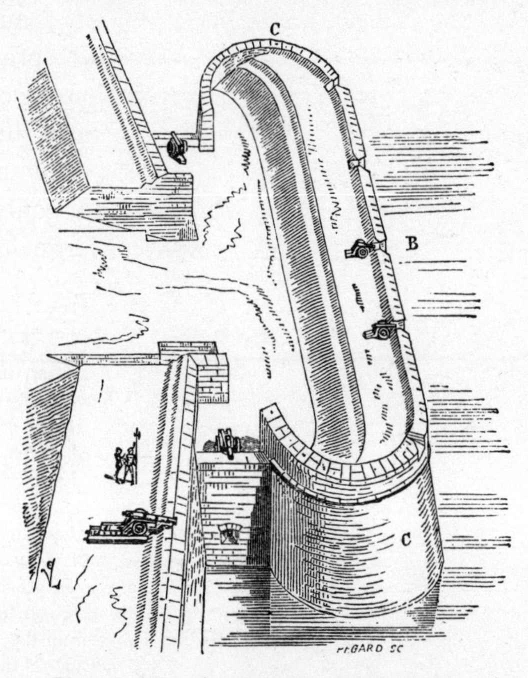

Fig. 130. View of an Orillon, or Oblong Bastion.

B. Straight face of Wall. C C. Rounded ends.

which were called orillons[o]. Those bastions enfiladed the ditches by means of masked pieces placed behind the

---

[o] The walls of the city of Narbonne, almost wholly rebuilt during the sixteenth century, and some ancient works in the fortifications of Rouen, Caen, &c., offered examples of defences constructed upon this principle.

orillons; but they only defended themselves on the face, they afforded no resistance to an oblique fire, and, above all, could not protect one another; their fire, in fact could not reach a breach battery (fig. 131) fixed at A, which would be exposed merely to the fire of the curtain. The attention of engineers was still so much directed to

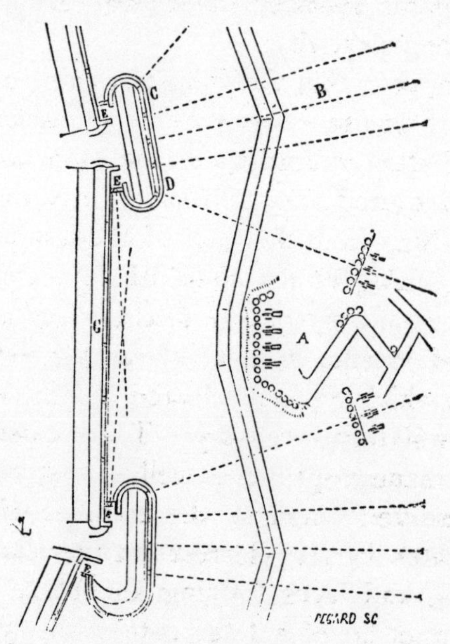

Fig. 131. Plan of Orillons.

A. Breaching Battery. B. Battery. C D. Straight Fronts. E E. Side Batteries. G. Breach.

the system of close defence, and they were so anxious to give to each part of a fortification an individual strength of its own (a principle inherited from the feudal military architecture of the Middle Ages, where each work, as we have proved, defended itself on its own account, as an isolated fort), that they looked upon the straight fronts, C D, intended for the destruction of the batteries placed at B, as necessary; reserving the fire at E, enfilading the curtains, merely for the moment when the enemy attempted the passage of the ditch, and made their assault at the breach effected at G. This last vestige of medieval tradition was not long in disappearing; and, from the date of the middle of the sixteenth century, a form was definitely adopted in bastions, which conferred on the fortification of strong places a force equal to the attack, up to the moment when siege artillery acquired an irresistible superiority.

It would appear that the Italian engineers, who at the close of the fifteenth century were so backward in the art of fortification, according to the evidence of Machiavelli, had acquired a certain superiority over those of France, resulting from the wars of the last years of that, and the beginning of the sixteenth, century. Between the years 1525 and 1530, San Michele fortified a portion of the city of Verona, and had already given to his bastions a form which was not adopted in France before the middle of the sixteenth century. There exists, however, a plan (in manuscript upon vellum) of the town of Troyes, preserved amongst the archives of that town, which indicates in the clearest manner large bastions with orillons, and faces forming an obtuse angle; and this plan cannot be of a later date than 1530, since it was made at the time when Francis I. had the fortifications of Troyes repaired, in 1524. Subjoined (fig. 132) is a fac-simile of one of the bastions shewn on this plan[p]. However this may be, the French engineers of the latter

---

[p] The ditch is a wet one. At A are shewn small masked batteries in two heights, held in reserve probably behind the covered flanks, B, constructed in rear of the orillons. Batteries, B, enfilade the front of the ancient towers, which were retained. It will be observed that the masonry which revets the bastion is thickest at the point and diminishes towards the orillons, that being a place where no breach could be effected; buttresses are placed as a stay to all the revetments, underneath the earth-works. This bastion is called *Boulevard de la porte Saint-Jacques*.

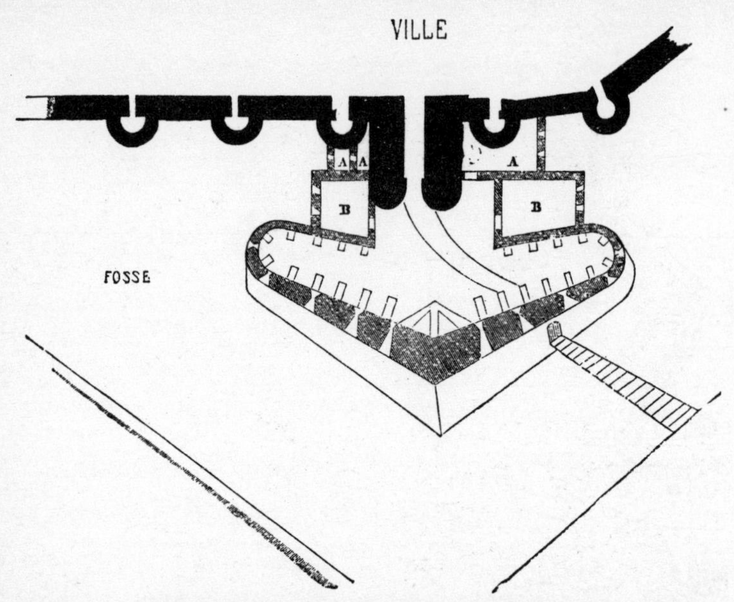

Fig. 132. Plan of one of the Bastions of Troyes.
A A. Small Masked Batteries. B B. Advanced Batteries.

half of the sixteenth century, abandoning the system of flat bastions, constructed them henceforward with two

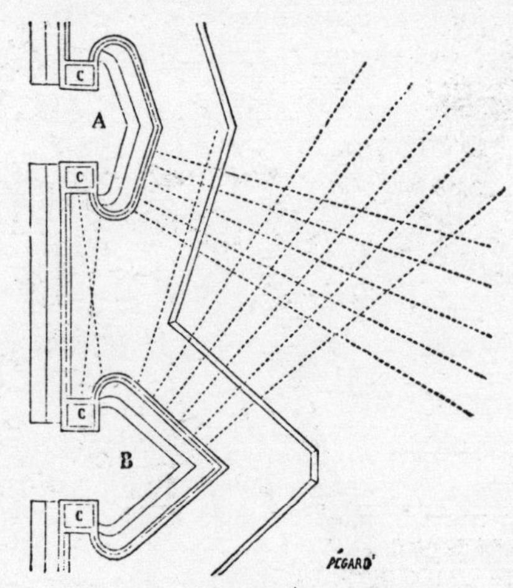

Fig. 133. Plans of Bastions.
A. Obtuse Angle. B. Acute Angle. C C. Casemated Batteries.

faces forming an obtuse angle, A (fig. 133), or forming

Fig. 134. View of Bastions attacked.

A. Bastions. B. Breaching Battery. C. Inner Rampart.

a right angle or an acute angle, B (fig. 133), in order to command the surrounding parts by cross-fires; keeping in reserve casemated batteries at C (sometimes these were in two heights), protected from the fire of the besiegers by the orillons, for the purpose of taking an assaulting column in flank, and almost in reverse, when the latter threw themselves into the breach. In the illustration we append (fig. 134), where this action is represented, it is easy to see the utility of flanks masked by orillons; one of the faces of the bastion, A, has been destroyed to allow of the establishment of the breach battery at B; but the pieces which arm the covered flanks of this bastion remain still intact, and can very materially damage the troops brought up to the assault, and throw them into disorder at the moment of their crossing the ditch, if, at the top of the breach, the attacking column are arrested in their progress by an internal rampart, C, thrown up in the rear of the curtain from one shoulder of the bastion to the other, and if this rampart is flanked by pieces of artillery. We have also shewn the bastion with a work thrown up across the gorge, the besieged foreseeing that they should not be able to defend it for any length of time. Instead of throwing up works across the gorges of bastions hurriedly, and often with insufficient means, the plan was adopted, from the close of the sixteenth century, of executing these works, in certain cases, in a permanent manner (fig. 135 [q]), or of detaching the bastions by

---

[q] *Delle Fortif.*, di Giov. Scala, al christ°. re di Francia ed i Navarra, Henrico IV., Roma, 1596. The figure here produced is entitled, "Piatta forma fortissima difesa e sicura con una gagliarda retirata dietro o attorno della gola." A, a rampart (says the legend) 50 ft. in thickness, of rear defence; B, a parapet 15 ft. thick and 4 ft. high; C, escarpment of the retirade, 14 ft. high; D, a space filled up solid, and slightly inclined towards the point G; H, flank-work, masked by the shoulder I; K, a parapet 24 ft. thick, raised 48 ft. above the ditch. (Scala here refers to the Roman foot = 11.72 inches Engl.)

sinking a ditch behind the gorge, leaving no communication with the body of the place, except by means of

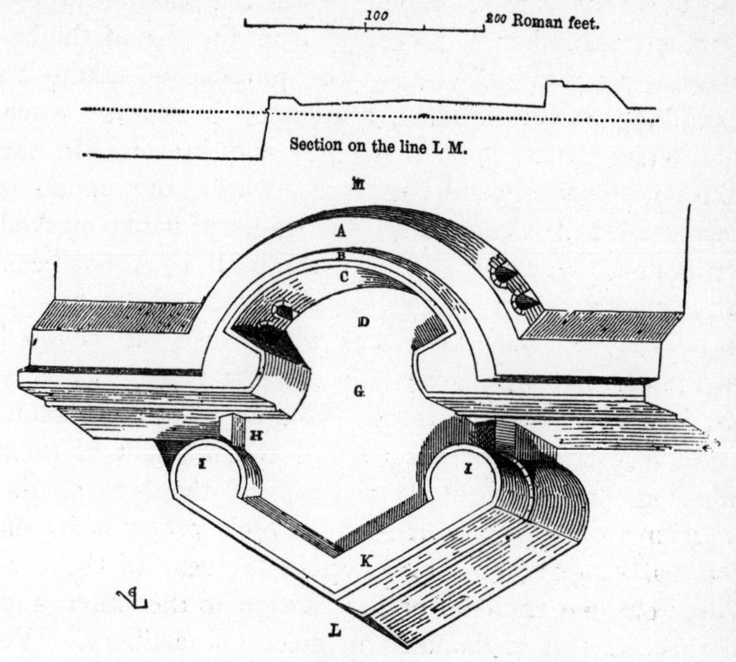

Fig. 135. Bastion isolated, with Inner Rampart.

A. Inner Rampart.
B. Parapet.
C. Escarpment.
D. Sloping Surface.
G. Lowest Point.
H. Flanking Battery.
I I. Masking Shoulders.
K. Parapet.
L M. Line of the Section.

drawbridges, or very narrow passages which could be easily barricaded (fig. 136[r]). By this means the taking of a bastion did not necessarily involve the immediate surrender of the body of the place; for it may readily be concluded that the besieging force endeavoured to

[r] *Delle Fortif.*, plate entitled "D'un buon modo da fabricare una piatta forma gagliarda et sicura, quantunque la sia disunita della cortina." X the legend describes as a rampart behind the curtain; C, a bridge which communicates from the city to the platform (bastion); D, solid earth-work; E, shoulders of the bastion; I, flanks to be made low enough to be covered by the shoulders E.

Scala gives, in his Treatise on fortifications, a large number of plans for bastions, some of them remarkable for the period.

effect a breach in the bastions rather than in the curtains, to avoid the direct effect of the masked batteries

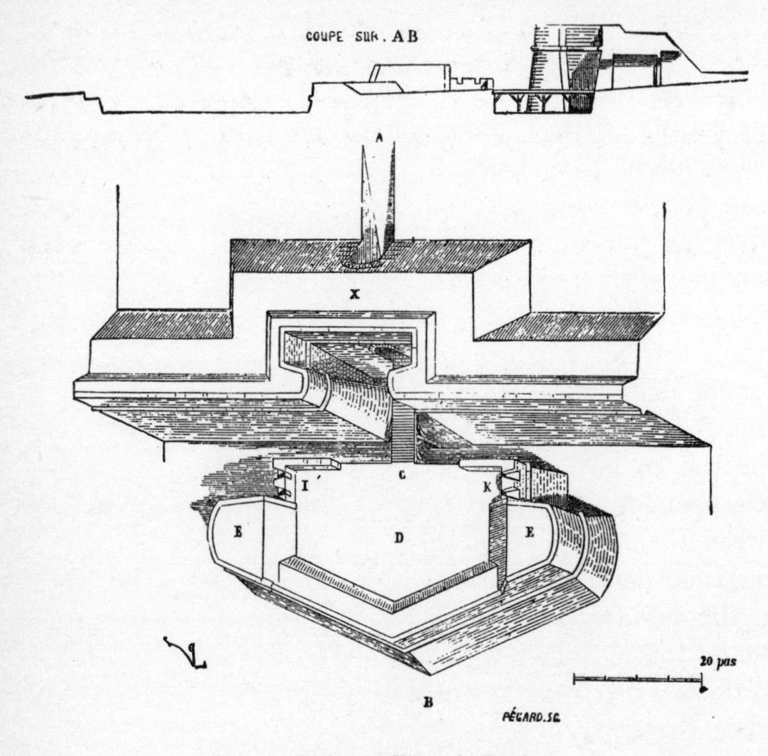

Fig. 136. Bird's-eye View of a Bastion.

A B. Line of the Section.
B. Parapet.
C. Bridge.
D. Solid Earth-works.
E E. Shoulders.
I K. Flanking Batteries.
X. Rampart behind the Bastion.

at the moment of assault. Seeing that the besiegers preferred to attack the bastions, with a view to breaching them and there making their assault, the engineers of the sixteenth century disposed the batteries masked by orillons in such a manner as to enfilade not only the curtain, but likewise the faces of the adjoining bastions. Thus, an assaulting column, whether it was thrown upon a bastion or on a curtain where a breach had been effected, was always met by a cross-fire; unless, indeed, the batteries masked behind the orillons had been silenced, previously to the assault, by ricochet shots or bombs.

How ingenious soever the expedients employed for defending the salient portions of the fortifications, and for cutting off their communication with the body of the place, might have been, no long time elapsed before it was discovered that these expedients had the defect of dividing the works, and of taking away the means of sending, with ease and rapidity, support to all the salient points of the defence, and that the advantages which resulted from their isolation were far from compensating for the dangers which this condition brought with it; so true is it that the simplest formulas are those which are the last to be adopted. The bastions, therefore, were left open at the gorge, but there were established between them, to protect their faces and in advance of the curtains, detached works which became of great utility in the defence, and which were frequently employed to hinder the approaches before feeble fronts or old walls; to these were given the names of *ravelins* or *demi-lunes*, when these works merely assumed the form of a small bastion, and of *ténailles* where two of those works were connected by a front. A (fig. 137) is a ravelin, and B a ténaille. Those works were already in use at the close of the sixteenth century, during the wars of religion; their slight elevation rendered it difficult to destroy them, while their horizontal fire produced a great effect.

It was also in the course of the sixteenth century that a decided batter was given to the revetments of the bastions and curtains, in order to neutralize the effect of the balls, which latter exerted, naturally, less action upon the wall-faces when they did not strike them at right angles. Before the invention of ordnance a slope was given only to the base of the revetments, in order

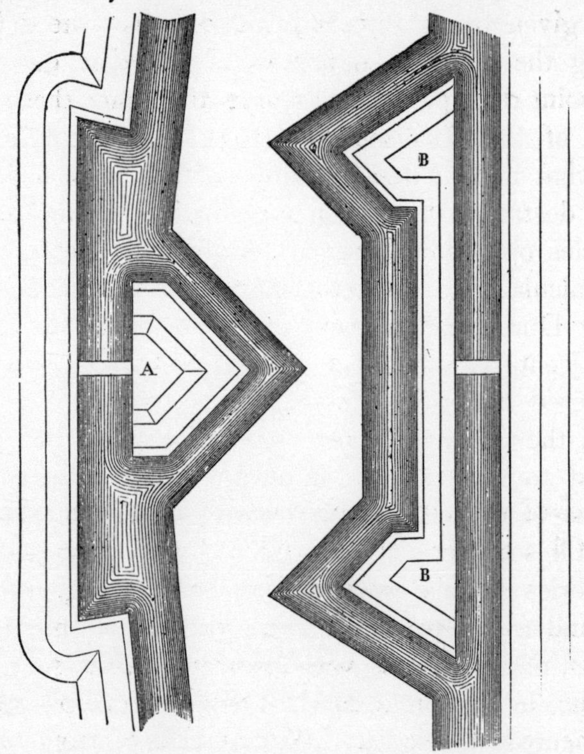

Fig. 137. Plans of a Ravelin, A, and two Tenailles, B.

to keep the assailants at some distance from the walls, and to place them vertically under the machicolation of the hoards; whilst, on the contrary, it was thought important to keep the walls vertical, to render it more difficult to scale them.

One very important detail appertaining to the defence of strong places must necessarily have engaged the attention of the constructors of fortresses when the use of cannon became general; we allude to embrasures. We have already seen how, in the fifteenth century, engineers had sought to mask the pieces placed in the interior of their defences as much as possible by various combinations more or less happy. The first embrasures, those given by us (figs. 89 and 98), had the defect of leaving the gunners so narrow a field that they could only point their pieces in a single direction; those of the castle of Schaffhausen (fig. 103), although offering a somewhat more extensive range of fire, must have been easily destroyed by the enemy's balls; the insignificant obstacles opposed to the artillery of the besiegers being only calculated to protect the gunners against musketry. Albert Durer had, so early as the first years of the sixteenth century, adopted a form of embrasure which, for uncovered batteries, offered signal advantages over the modes then commonly received. Those embrasures, as applied to the barbette batteries of the bastions and curtains of the city of Nuremberg, and which any one may still see there, are reproduced with their necessary accessories in his work[s]. We subjoin (fig. 138) the plan and (fig. 139) the section of one of them. The parapet, of a thickness varying from three to four yards, presents, in section, a curve intended to throw upwards the enemy's projectiles. A mantelet of stout wooden planks, revolving upon a horizontal axis, and forming an angle with the horizon, which was only elevated sufficiently to afford a passage for the muzzle of the gun and allow the piece to be pointed, offered no resistance to the balls of the besiegers, and sent them ricochetting over the heads of the gunners. This system does not appear to have been adopted in France, where the parapets from an early period had been covered with earth and grass, having embrasures furnished with fascines while

[s] "Alb. Dureri pictoris et architecti præstantiss. de urb. arcibus, castellisque condendis ac muniendis, rationes aliquot, præsenti bellorum necess. accomm.: nunc recens è ling. German. in Latinam traductæ." (Parisiis . . . 1535.)

the siege lasted. Besides the example already given, the parapets of the curtains and bastions of the city of

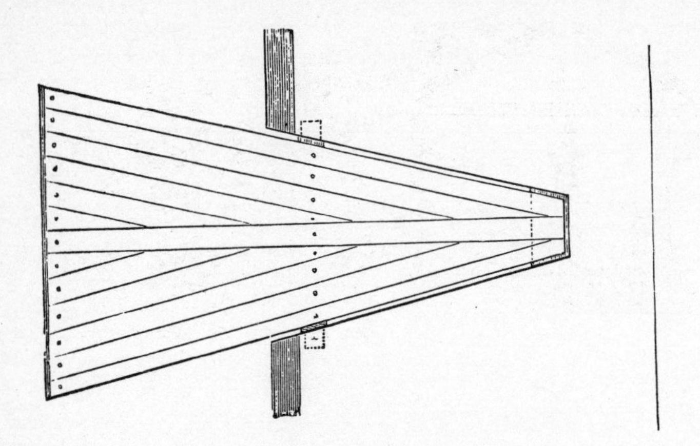

Fig. 138. Plan of an Embrasure at Nuremberg.

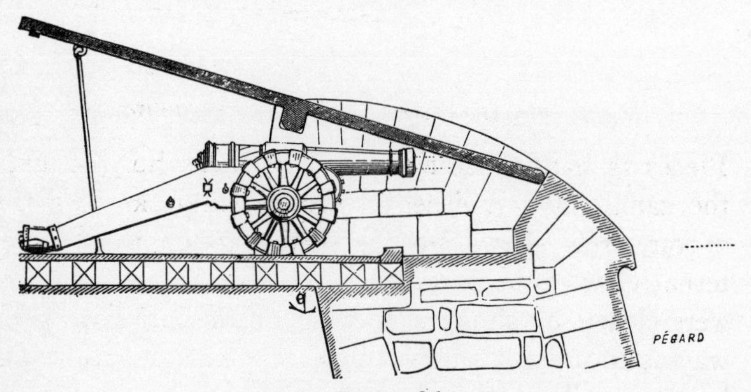

Fig. 139. Section of the same.

Nuremberg, erected by Albert Durer, present, throughout a large portion of their extent, and principally on the side where the fortifications are accessible, a remarkable arrangement which we here (fig. 140) reproduce. These parapets, pierced with embrasures for cannon, are surmounted by timber hoards (*hourdés*), or filled in with brick and mortar, like the old English half-timbered houses: in those hoards, arquebusiers and even archers (who were still employed at this period) might be placed.

Fig. 140. View of the Parapet at Nuremberg, with the Hoarding.

Pieces in battery were covered by these hoards, just in the same way as pieces in the 'tween-decks of a man-of-war, as is shewn by the section given with the external view of the parapets. The crenelles of the hoards were closed by shutters opening on the inside, in such a way as to present an obstacle to the balls or arrows fired by the assailants placed on the top of the glacis.

We have sometimes seen in France the embrasures of uncovered batteries presenting externally a series of broken faces intended to stop the balls and bullets of the enemy (fig. 141), and to hinder them from sliding, as they would, along an unbroken splay to the mouth of the cannon. The embrasures of covered batteries, however, long retained their original form, that is to say, they consisted only of a round or oval aperture and a sight-hole, nor was it until the close of the sixteenth century that they were made to widen backward from

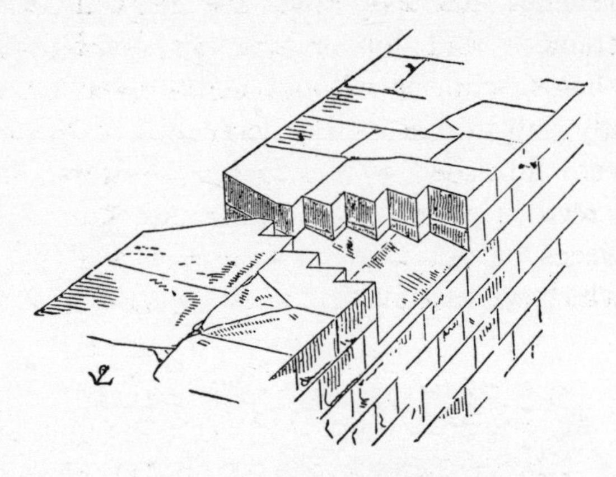

Fig. 141. Embrasure with Redents.

beneath an arch (fig. 142). Our artillery-men soon remarked that the narrow part of the embrasure ought not

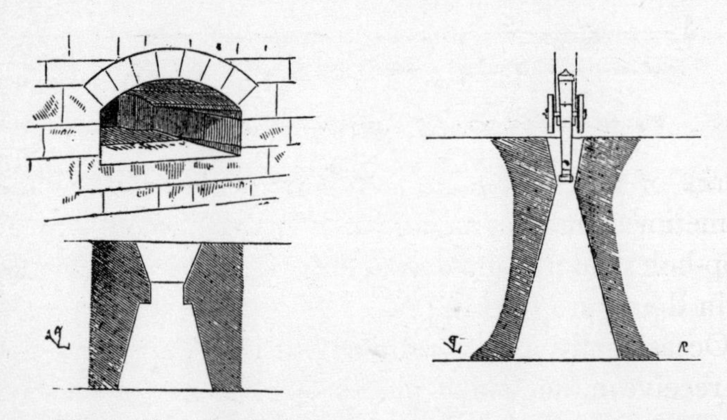

Fig. 142. Plan and View of an Embrasure. Fig. 143. Plan of another Embrasure.

to reach to the middle of the thickness of the walls of casemates, for these walls being from six to seven yards thick, the narrow part of the embrasure, which lay beyond the mouth of the cannon, was soon shattered by the wind of the piece; they therefore gave the embrasures of casemated batteries the form in plan represented by the figure 143.

In crenelles and loop-holes the original forms were long retained; but for arrow-loops (*archères*) simple conical holes, with or without sights over them, were frequently substituted[t]. The crenelles of the covered-ways were furnished with hanging shutters, having a hole pierced in them, and adapted either to the fire of small pieces or of arquebuses, as indicated by the example which we subjoin (fig. 144), copied from the crest-

Fig. 144. Covered-way, with Crenelles, Loopholes, and Shutters.

works of the curtains at Nuremberg (fifteenth century). Sometimes embrasures for cannon were accompanied by loop-holes, lateral and descending, for musketry, arranged as in the figure (fig. 145[u]).

Occasionally, also, certain embrasures were constructed to receive either small pieces of ordnance, such as falconels, or those large rampart arquebuses which may still be seen in the French and German museums, and

---

[t] The name of crenelle is at present given to the small embrasures pierced in parapets for musketry, and similar enough in form to the ancient arrow-loops; whilst anciently, the name of crenelle (or *créneau*) was given to the square open space left between the two merlons of a parapet.

[u] Bastions of the city of Nuremberg of the close of the fifteenth century.

of which there is a great number in the arsenal at Basle. As an example of these latter embrasures we may give

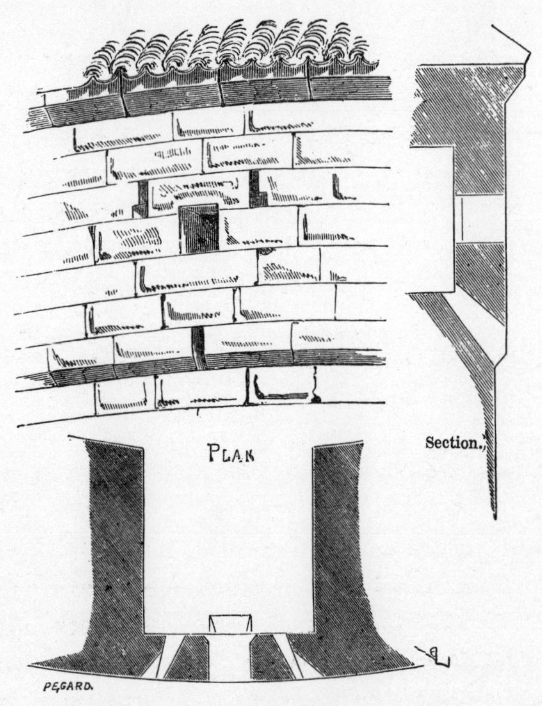

Fig. 145. Elevation, Section, and Plan of an Embrasure, with Loopholes for Musketry.

those of the advanced work of the Laufer gate, at Nuremberg, which are very curious and worth studying. This outwork, perfectly intact, and which has preserved the greater part of its accessories of defence, dates from the middle of the fifteenth century. The void of the embrasures (fig. 146) is a vertical oblong in form, and facilitates therefore the plunging fire of the pieces, commanding the bottom of the ditch as well as the glacis. This void, or crenelle, is furnished with a stout wooden cylinder placed upright with hoops and pivots of iron. The cylinder is pierced from side to side for a portion of

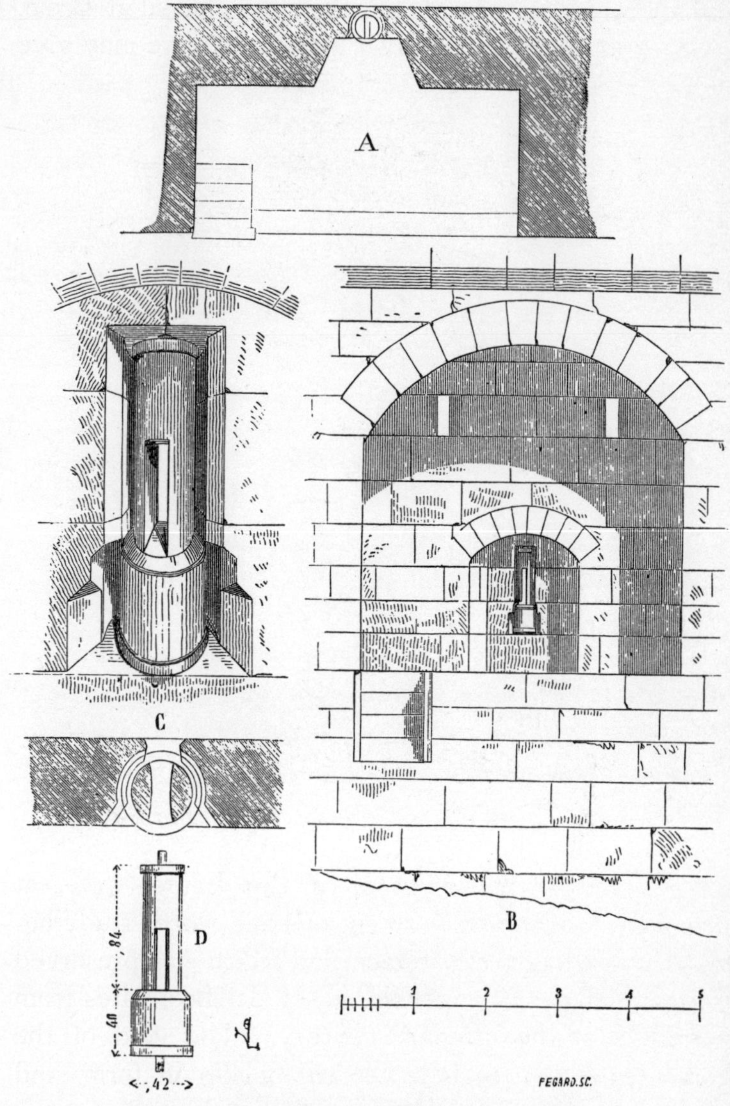

Fig. 146. Embrasure of the Laufer Gate at Nuremberg.

A. Plan. B. Internal Elevation. C. Plan of Loophole, with Turning-post. D. Form of the Turning-post.

its height by an oblong aperture about $4\frac{1}{2}$ inches wide by $9\frac{1}{4}$ inches high, which is just enough to allow of the passage of the ball of the rampart pieces. When the piece was discharged, by turning the cylinder on its axis, the men placed in the embrasure were completely masked. A shews the general plan of the embrasure, B its internal elevation, C the plan and elevation of the crenelle with the revolving cylinder, and D the geometrical form of the cylinder with its dimensions (in parts of the French centimètre). The advanced-work of one of the gates of the city of Basle still retains its crenelles (or loops) thus furnished with wooden cylinders, longitudinally pierced, for passing the muzzles of hand-arquebuses through.

But it must be admitted that, in the presence of artillery, all those defensive expedients must have soon appeared insufficient, and rather an incumbrance than of any real service: neglecting therefore such precautions in fortresses, too convinced perhaps of their little utility, engineers contented themselves with embrasures of the simplest form, such as is shewn (fig. 143), consisting of an open crenelle forming an angle more or less acute, leaving barely space enough for the passage of the piece; and it was only when the siege took place that preservative means were taken to protect those placed in the casemates and uncovered batteries. After having attached too great an importance to those details of defence, when the use of ordnance had been the means of radically altering the art of medieval fortification, it may be that their importance has been under-estimated since the seventeenth century. It is certain that, against the shock of artillery, it is useless to think of opposing any obstacles but those which shall be at once of great power, and yet simple enough not to retard the service of the guns while admitting of their being promptly and easily replaced.

From the moment when bastions assumed definitely a new form, the system of attack, as well as the system of defence, became completely changed. The approaches had to be skilfully planned, for otherwise the cross-fire of the faces of bastions enfiladed the trenches and took the siege-batteries obliquely. The trenches had to be commenced at a great distance from the body of the place; distant batteries had to be established, to destroy the parapets of bastions whose fire might annihilate the works of the sappers; in order thus by degrees, and under cover of the works thrown up, to arrive at the back of the ditch, whilst at intervals *places d'armes* protected the batteries and trenches against night sorties by the besieged; until, finally, the last battery was established to effect a breach. It is needless to say that even previously to the time when the art of fortification had become subject to regular formulas, before the times of such men as Errard de Bar-le-Duc, Antoine Deville, Pagan and Vauban, engineers had been driven to abandon the last traditions of the Middle Ages. But starting with the rule, *that whatever defends ought to be defended*, impediments were so multiplied, so many separate works and commanding positions were established, the defences were encumbered with so many details, and such care was taken to detach them one from the other, that they became for the most part useless and even hurtful when the siege took place; and the garrison, ever sure of finding a second line of defence when the first was destroyed, and a third after the second, defended themselves feebly in one after the other, always trusting to the last to make a stand.

Machiavelli, with the practical sense which is his characteristic, had always foreseen in his day the danger of those complications in the construction of works of defence; for in his "Treatise on the Art of War," book vii., he says,—

"And here I ought to give an advice: 1stly, to those who have the charge of defending a city, namely, never to erect bastions detached from the walls; 2ndly, to those who are constructing a fortress, and that is, not to establish within its circuit fortifications which may serve as a retreat to troops who have been driven back from the first line of entrenchments. The reason for my first advice is this: we should always avoid a failure at the beginning, for we thus beget a distrust of all our future plans, and fill those who have embraced our cause with apprehension. You will not be able to provide against these mishaps by erecting bastions beyond the walls. As they will be constantly exposed to the whole fury of the artillery, and as at the present day such fortifications cannot be defended for any length of time, you will end by losing them, and will thus have prepared the cause of your ruin. When the Genoese revolted against Louis XII., king of France, they built in this manner some bastions on the surrounding hills; and the taking of those bastions, which were carried in a few days, brought with it the loss of the city. As regards my second proposition, I maintain that there is no greater danger for a fortress than rear-fortifications whither troops can retire in case of a reverse; for once the soldier knows that he has a secure retreat after he has abandoned the first post, he does in fact abandon it, and so causes the loss of the entire fortress. We have a very recent example of this in the taking of the fortress of Forli, defended by the Countess Catherine against Cæsar Borgia, son of Pope Alexander VI., who came to attack it with the army of the King of France. This place was full of fortifications where retreat after retreat might be secured. There was, first of all, the citadel, separated from the fortress by a ditch which was crossed by means of a drawbridge, and this fortress was divided into three quarters, each separated from the other by ditches filled with water, and drawbridges. Borgia, having attacked one of those quarters with his artillery, effected a breach in the wall, which breach M. de Casal, Commandant of Forli, did not attempt to defend. He thought he might abandon this breach and retire upon the other quarters; but Borgia, once master of this portion of the fortress, was soon the master of the whole, because he seized upon the bridges which separated the different quarters. Thus was taken a place which was then considered almost impregnable, and the loss of which was due to two principal errors on the part of the engineer who had constructed it: 1stly, he had too much multiplied the defences; 2ndly, he had not left to each quarter the command of its own bridges[1]."

Artillery had changed the moral conditions of the defence quite as much as the material conditions: just as it was good, in the thirteenth century, to multiply impediments, to erect fort after fort, to break up the defences, because both attack and defence were made foot by foot and hand to hand; just in the same way was it dangerous, when the powerfully destructive effects of artillery had been brought to bear, to interrupt the communications, to encumber the defences; for the cannon destroyed those complicated works or rendered them useless, and burying the defenders beneath their ruins, demoralized the defending force, and deprived it of the means of united action.

It had been ascertained already, in the system of fortification prevailing before the introduction of ordnance, that the extreme sub-division of the defences rendered the task of command a difficult one for the commandant of a fortress, or even for the captain of a post; in detached works, such as towers, donjons, or gate-houses, the necessity had been felt, as early as the eleventh and twelfth centuries, of opening conduits or traps in the walls or through the vaulting, to serve as speaking tubes by which the commanding officer of the

---

[1] Complete Works of N. Machiavelli. See the castle ot Milan (fig. 121), which offers examples of all the faults pointed out by Machiavelli.

post, while placed at the most favourable point for obtaining a view of the operations on the outside, could transmit his orders to each story and portion of the works. But when the roar of artillery came to be superadded to these material difficulties, it will readily be understood that those means became altogether insufficient; the use of cannon, therefore, necessarily induced in the construction of fortifications a greater breadth of arrangement, obliging both the besieged and besieging forces to abandon altogether a war of details.

The method which consisted in fortifying strong places outside the line of the old walls had its inconveniences: the besieging force were able to operate at one and the same time against the two lines, the second being higher than the first; they thus destroyed the two defences, or at least, overthrowing the first, they dismantled the second, shattered its merlons into fragments, and dismounted simultaneously the lower and the upper batteries (see fig. 116). If they succeeded in carrying the front line of defences, they might still be stopped for some little time by the escarpment of the old wall; but this latter, being deprived of its barbette batteries, became no more than a passive defence which could be blown up without danger, or without requiring the assailants to cover themselves. For this reason Machiavelli had already in his day recommended that permanent ramparts, with a ditch, should be erected in the rear of the old walls of cities. Allowing, therefore, the ancient walls to remain as a first obstacle to resist a *coup-de-main*, or to arrest the progress of the enemy for some little time, and abandoning the use of external boulevards and of salient works which were exposed to the converging fire of the siege-batteries and were soon destroyed, the constructors of fortifications erected—in the rear of the ancient lines, which, from their weakness, would naturally be selected by the enemy for their point of attack—bastioned ramparts, forming a permanent work, analogous to the temporary one which we have shewn at fig. 109. It was upon this principle that a portion of the city of Metz had been fortified (after the raising of the siege undertaken by the imperial army, towards the close of the sixteenth century), on the side of the Sainte-Barbe gate (fig. 147[y]). Here the ancient walls, A, with their lists, are left just as they were; barbette batteries only being established in the ancient lists, B. The enemy, having made a breach in the front, C D, which was actually the weakest inasmuch as it was not flanked, and having crossed the ditch and arrived at the *place d'armes*, E, was exposed to the fire of the half-bastions, F G, and to both a front and cross-fire. From the outside, this rampart, being lower than the old wall, remained masked, intact; its flanks with orillons contained a covered and uncovered battery, enfilading the ditch.

The great merit of the engineers of the seventeenth century, and of Vauban especially, consists in their having arranged the defences in such a manner as to converge upon the first front attacked and destroyed by the enemy the fire of a great number of pieces of artillery, and thus to change, at the moment the assault took place, the relations between the besieging and besieged armies; and in their having simplified the art of fortification and done away with a vast number of detached works and details of defence, which are very ingenious on paper, but which are only impediments during a siege, and impediments of a very costly kind. It was

---

[y] *Topog. de la Gaule*, Mérian; *Topog. de la France*, Bibl. Imp.

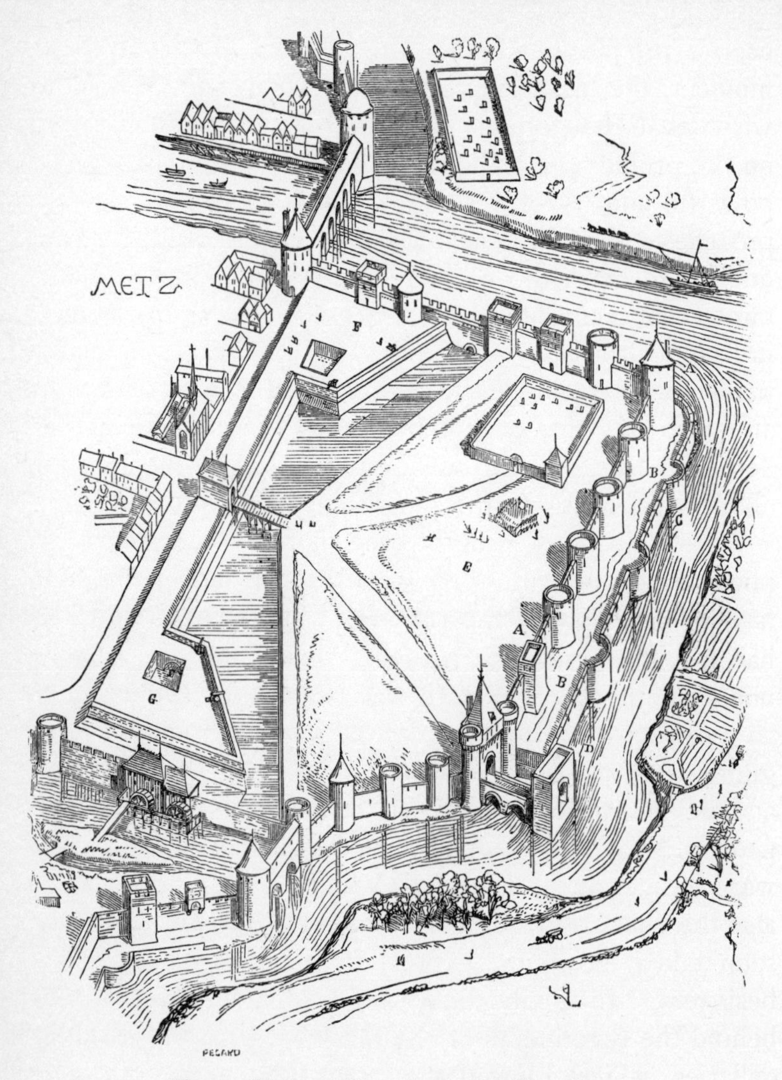

Fig. 147. Bird's-eye View of part of the City of Metz.

A. Ancient Walls. B B. The Lists. C D. Front of old Walls.
E. Place d'Armes. F G. Half-bastions.

thus that, by degrees, the superficies of bastions was enlarged; that orillons of small diameter which, destroyed by the artillery of the besieging force, rendered useless the pieces intended to enfilade the ditches at the moment the assault was delivered, were done away with; that the greatest possible attention was bestowed on the profiles (or sections), these being one of the most powerful means of retarding the progress of the approaches; that the glacis was thrown up in advance of ditches, to mask the revetment of the bastions and curtains; that a considerable width was given to the ditches in front of the *fausses braies;* that stone revetments for parapets were replaced by embankments of sodded earth, and that the gateways and gate-houses were masked, defended by advanced works, and flanked, instead of allowing their strength to consist in themselves alone.

A new means of rapidly destroying ramparts was applied at the beginning of the sixteenth century: after having undermined the revetments of the defences, as had been practised from time immemorial, instead of underpinning them with shores which were then set fire to, pockets (*fourneaux*) were made, and charged with gunpowder, and considerable portions of the earth-works and revetments were thus blown up. This terrible expedient, which had already been employed in the Italian wars, besides opening large breaches for the assailants, had also the effect of demoralising the garrison. Means were soon taken, however, to neutralize those works of the besiegers. In fortifications where the ditches were dry, behind the revetments of the ramparts were run vaulted galleries, which allowed the garrison to resist the construction of those *fourneaux de mine* (fig. 148 [z]); or, at intervals along the solid earth-works of the parapets, permanent wells were sunk, in order therefrom to push forward counter-mine galleries while the siege was going

[z] *Della Fortif. della Citta* di M. Girol. Maggi, e del cap. Jacom. Castriotto, ingeniero del christ°. re di Francia, 1553.

on, and after the engineers of the besieged had succeeded in ascertaining the direction of the galleries of

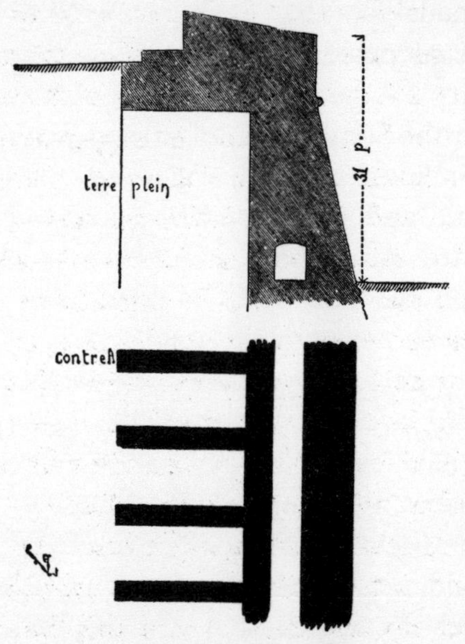

Fig. 148. Plan of Vaulted Gallery.

the enemy's mines; which direction could be made out by carefully observing, at the bottom of those wells, the noise made by the sap. Occasionally, also, countermine galleries were run under the covered-way or beneath the glacis; but it does not appear that these latter means were applied in any regular manner until the adoption of the system of modern fortification.

It was only by slow degrees, and as the result of numberless experiments, that scientific formulas could be arrived at in the construction of defensive works. During the course of the sixteenth century we find the germs of almost all the systems subsequently adopted, but there is no general method, no unity of plan; the monarchical power, which is one in its essence, could alone lead to any definite result: and it is interesting to observe how the art of fortification, as applied to artillery, follows step by step the preponderance of the royal over the feudal power. It is not until the commencement of the seventeenth century, after the religious wars under Henry IV. and Louis XIII., that the works connected with the fortification of strong-places are planned after certain fixed rules, based upon a long course of observation; and that the last remains of the ancient traditions are abandoned, and formulas adopted, established upon the new bases of calculation. Thenceforward it became the unceasing endeavour of all engineers to find a solution for the problem, *To see the besieging force without being seen, while obtaining a cross and defile fire.* The exact solution of this problem would render a fortification perfect and impregnable; but that solution, in our opinion at least, has yet to be discovered.

We should not be able, without entering into long details which do not come within the limits of our subject, to describe the various experimental efforts made since the beginning of the sixteenth century to raise the art of fortification up to the point at which Vauban has left it. We shall merely give, in order to furnish some idea of the new principles upon which modern engineers were about to establish their systems, the first figure of the Treatise of the Chevalier De Ville[a].

"The hexagon," says that author, "is the first figure that can be fortified, the bastion remaining at right angles; which is the reason of our commencing with it, of which, having given the method, it may be applied in like manner to all the other regular figures (fig. 149). Firstly, let a regular figure be constructed, that is to say, one having the sides and angles equal, of as many sides as it is required that the figure shall have bastions . . . . In this figure we have drawn one half of the

[a] *Les Fortifications* du Chevalier Antoine De Ville, 1640, chap. viii.

hexagon, on which having shewn how to make a bastion, the same can be done with all the other angles. Let it be the angle

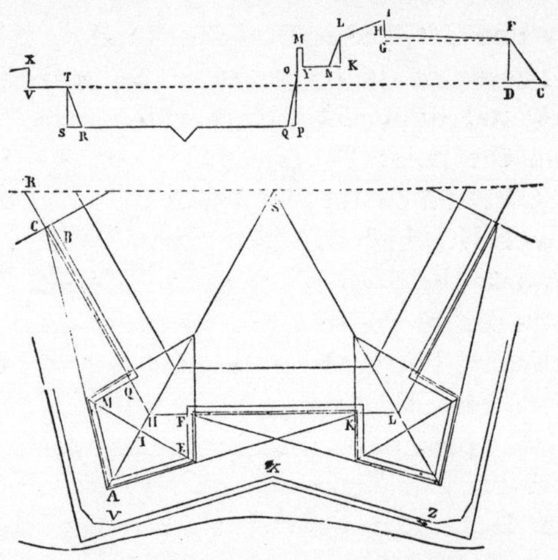

Figs. 149 and 150. Plan and Section of a Bastion according to De Ville.

R H L of the hexagon, on which it is required to make a bastion. Divide one of the sides, H L, into three equal parts, and each of these parts into two, which are H F, and set off H Q equal to H F . . . . these are the demi-gorges of the bastions; and on the points, F and Q, erect, perpendicularly, the flanks F E, Q M, equal to the demi-gorges; from one extremity of the flank to the other draw the line M E and produce the semi-diameter S H . . ., and make I A equal to I E; then draw A E, A M, which will form the rectangular bastion Q M A E F, and will provide as much defence to the curtain as is possible; and where the said curtain begins, may be known by producing the faces A E, A M, until they meet the said curtain at C and at K; the line of defence will be A C . . . .

"It will be observed that this method will not answer for places having less than six bastions, because the flanks and gorges being of the same length, the bastion forms an acute angle. As for the other parts, make the line of the ditch V X, X Z, parallel to the face of the bastion, at a distance from the latter equal to the length of the flank."

De Ville admits the orillons or shoulders to the flanks of bastions; but he prefers rectangular to circular orillons. He gives with the plan (fig. 149) the profile of the fortification (fig. 150).

"Draw any line C V," adds De Ville, "and on this take C D, equal to five paces; on the point D erect the perpendicular D F, equal to C D, and draw C F, which will be the slope of the parapet: from the point F draw F G, equal to fifteen paces, parallel with C V, and on the point G raise G H, equal to one pace, and draw F H, which will be the plane of the rampart with its incline towards the body of the place. Make H I four feet, and G L five paces for the thickness of the parapet; K L must be drawn vertically, but K should be placed at two paces above the line C V; afterwards draw K N, the batter (or talus) of the parapet, N Y, the parapet-way, shall be about two paces, and M less than a half-pace in thickness, and its height, M Y, will be seven or eight feet; then let M P be drawn perpendicular to C V, so that it shall be five paces under O, that is to say, that depth under the ground-level, and this will be the depth of the ditch. P Q is the batter or outward slope of the wall, which should be a pace and a-half, and O will be the stringcourse (*cordon*), a little over the esplanade: the width of the ditch, Q R, in large fortifications should be twenty-six paces, and in others twenty-one paces. Let R S, the slope of the counterscarp, be two and a-half paces, and its height, S T, five paces; the corridor (covert-way), T V, which should be placed on the line C V, shall be five to six paces in width, the esplanade (the glacis) shall be one pace and a-half above the corridor V X, and said esplanade shall slope down to the country some fifteen or twenty paces... make the profile thereof... of which there are divers kinds...; the paces being equal to five *pieds de roy*."

De Ville recommends *fausses braies* in advance of the ramparts as greatly increasing the strength of fortifications, for the reason that, being masked by the profile of the covered-way, they retard the establishment of breach-batteries and command the points where the trenches debouch upon the ditch: he considers they should be made of earth, and in the manner indicated by the profile at A (fig. 151).

It was then with fortification as with every other branch of the art of architecture,—formulas had become

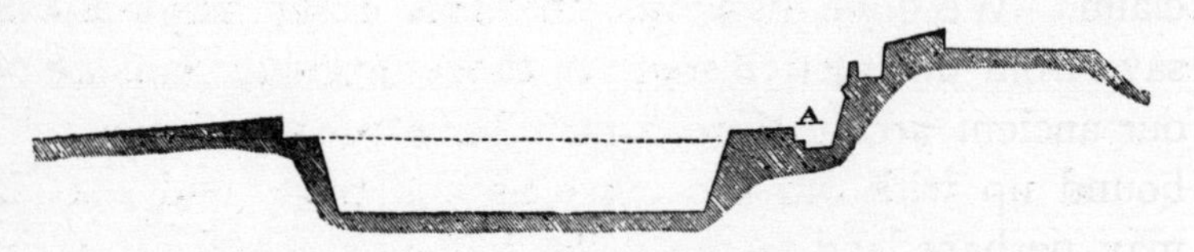

Fig. 151. Section of Ditch with False Braie, according to De Ville.

the rage, and each engineer brought forward his own system : if we have spoken of the Chevalier De Ville, it is because his methods are practical and the result of experience. But Vauban discovered that the bastions constructed by the engineers who had preceded him were too small, their flanks too short and too weak, the demi-gorges too narrow, the alignement of the ditches badly set out, and the covered-ways too limited in width, the *places d'armes* small and the external works insufficient. It is to him and to M. de Coëhorn that we owe systems of fortification very far superior to those which had preceded them. Nevertheless, according to the admission of those celebrated men themselves, in spite of all their efforts, the attack was still superior to the defence.

The study of the works executed during many centuries by several generations of men to defend their lives, their liberty, and their fortunes, is certainly one of the most attractive that can be pursued ; it is likewise, perhaps, one of the most useful. This study is connected with the successive developments of the national civilization and character, and it must be allowed that no country in Europe presents a more considerable series of permanent military works than France. We have been able, in a work so limited as this is, merely to give a very summary idea of a subject so vast, and which would require, on the part of any one desirous of entering into it fully, an extent of information to which we can lay no claim. We hope, however, that this Essay may help to save from destruction some of those precious remains of our ancient architecture, which have been so intimately bound up with our existence as a nation; and that it may perhaps lead to the collection, in a complete work, of the numerous fragments of military architecture which cover the soil of France, and which the hands of men and individual interests, even more than the ravages of time, are every day destroying. It would be worthy of an enlightened Government like ours to undertake this task, far beyond the powers of a single man. In addition to the archæological interest which would attach to a work of this importance, it would read us more than one lesson; we should therein obtain a knowledge of the resources of a country which we love, because she is our own, and, better still, because she has struggled always after a national unity, and because her energy has always made her rise superior to her reverses.

INDEX.

(*The Asterisks refer to Engravings.*)

AGDE, walls of the town of, 3.
Aigues-Mortes, towers, 122
  *Plan of the town, 123.
  *ramparts, 132, 145.
Aiguillön, the siege of, 155—157.
Aix road, the, 222.
Albigenses, 33, 43.
Alexander VI., Pope, 254.
Alure, or *allure*, the walk behind the parapet.
  *At Carcassonne, 11.
  *—— Plan of, 66.
  *At Coucy, 110.
  *—— Plan of, 148.
Andelys (les), 80, 81, 84.
  Andely, grand, 84.
  ——— Isle of, 92.
  ——— Petit, 84, 92, 93.
Anglo-Norman feudalism, 78.
Antoninus, the Column of, 5, 6.
Arbalétriers, the Corporation of, 165.
**Archers and bowmen*, 173—175.
Arno, the river, 219.
Arragon, King of, 47.
Arras, bastille of, 217.
Arrow-loops (*archères*), 249.
Artillery, introduction of, 166.
  Early use of, 168.
  The English expelled by improvements in, 169.
  Further improvements, 170, 171.
  *Of the fifteenth century, 172.
  Towers altered to receive, 177.
  *Embrasures for, 181.
  Modifications of walls, &c., 182.
  *Plans of towers constructed for, 184, 185.
  Castles cannot resist, 187.
  Modes of defence against, 202—206.
  Description of French and English in the sixteenth century, *note*, 207.
  Increasing power of, 215, 234.
Artillery, irresistible, 237.
  *Improved embrasures for, 248.
  Changes the conditions of the defence, 255, 256, 262.
Artillery Museum of Paris, 172.
Aubenton, the town of, 136.
  *The gate of, 135, 136.
  Siege of, 125.
Aude, the river, 46, 47, 49, 51, 52, 54, 58.
Augsburg, *fortifications of, 229—232.
Autun, Roman city of, 15.
Auxerre, Roman city of, 15.
Avaricum (Bourges), 5.
Avignon, *walls of, with details, 144—150.
  *Inside view of towers at, 146.
  *—— Plans of the same, 147.
  *Perspective view of the interior, 149.
  *Palace of the Pope, 150.

Bacon (a brigand), 162.
Barbacane, La, 46.
Barbette battery, or *batteries*, 184, 186, 199, 257.
**Barbicans*, or *outworks*, 37, 51, 52, 75, 132.
Bartizan (*échauguette*), a watch-turret, as at Carcassonne, 55.
  Provins, 104.
  Schaffhausen, 196.
Basle, or Bâle, 217.
  *Wooden cylinders at, 252.
  The arsenal of, 171.
Bastille, or *bastilles*, Roman, 18.
  *Section of, 20, 29, 68, 167, 203, 217, 224.
Bastille of S. Antoine, 75, 76.
Bastions, *plans of, 119, 120.
  *At Schaffhausen, 194.
  *At Verona, 224.
  *At Augsburg, 230, 231.

Bastions at Frankfort-on-the-Maine, 234.
*Plans of, 238.
*Attacked, 239.
*At Troyes, 238.
*Isolated, 241.
*Bird's-eye view, 242.
*Plan of, according to Deville, 262.
*At Carcassonne, 66.
With wooden door, 67.
Batteries, 211, 212.
**Battering-rams*, 7, 25, 33, 36, 42, 43, 119, 203, 204, 207.
**Beak*, *horn*, or *pointed bastion*, 121, 122, 127.
*Plans of, 122.
Beaucaire, siege of, 26, 43.
Beaugency-sur-Loire, 98.
Bernard Hugon de Serre-Longue, 38.
Bernières, peninsula of, 92.
Béziers, early fortifications of, 3.
Bibl. de l'Ecole des Charts, 166.
Bicoque, a, 69.
Blanche, Queen, 37, 38.
Blaye, *Defenses de la Ville de*, 226.
Boccanegra, the Genoese, 145.
Bombard, a great gun, short and thick, nearly equivalent to the modern bomb, used at the siege of Calais, 158.
Bonaguil, the castle of, 177—181.
*Plan of, 178.
*Bird's-eye view, 180.
*Embrasure, 181.
Bonaparte, L. Napoleon, The Past and Future of Artillery, 166.
Essay on the Influence of Fire-arms, 201.
Bonnières-on-the-Seine, 81.
Bordeaux, Roman fortifications of, 15.
Bastille of, 217.
Bosson (*langue d'oc*), a battering-ram, 25, 27, 33, 43.
Boulevards, or *bastions*, 214, 215, 226, 227.
De la Porte St. Jaques, 237.
Boulevert bastille, or *bastide*, 201.
Boulogne, the gates of, 158.
Bourbon-l'Archambaut, 77
Bourgeois, le, 167.
Bourges, siege of, 5.
Bourgogne, le Duc de, 229.
Boussac, town of, 77.
Boutavant, fort of, 84.
Boves, siege of, 34, 35.
Brabançons, the, 153, 160.
Braie, or external ditch, 201, 202.
*Section of a false braie, 264.
Brattish (breast-work), 40, 45, 58, 125, 128, 130, 131, 137, 138.
Bridge of boats, with turrets on it, 92.
Brittany, hôtel of the Dukes of, 73.
Bureau, the brothers, 166, 168.

Cables used by the Romans to form mats for defence of walls, 19.
Caen, siege of, 167; orillons at, 235.
Cæsar, the time of, 4, 5, 18, 22, 33.
Cæsar's Commentaries, 4.
Cæsar Borgia, 254.
Cahors, Roman fortifications of, 15.
Calabres, battering machines, 36.
Calais, siege of, 155, 157—159.
Cambray, defence of, 166.
Cannon, 171, 172, 176, 177, 181.
Varieties of, in use in England and France in the sixteenth century, 207, *note*.
Capendu, Raymond de, 42.
Carcassonne, early walls of, 3.
Of the Visigoths, 10.
*Plan of the Visigoth, or Roman walls, 17.
Siege of, in 1240, 37—42.
*Wood-work to defend a breach, 41.
*Plan as fortified by S. Louis, 48.
Fortifications of, 47—59.
*Plan and *bird's-eye view, 56, 57.
Castle of the twelfth century, 8.
*Pointed towers, or horns at, 121.
*Plan of the Narbonne gate, 126, 128, 131.
*Elevation of, 129.
Bishop's palace at, 51.
*Towers at, 11, 12, 13.
Carlovingian period, 24.
Casal, M. de, 254.

Casemated battery, a vaulted chamber with embrasures for cannon, 185, 195, 251.
Castellain of Cambrai, 114.
Catapult (*trébuchet*), 34, 36, 43, 63, 64, 66, 94, 115, 136.
Catherine, the Countess, 254.
Cats (*chas*), 34, 36, 51, 62, 94, 150.
Celts, fortifications of the, 4.
Chabannes, Jacques de, 167.
Chaillot, slopes of, 73.
Charlemagne, 24.
Charles the Bold, 201.
——V. (of France), 75, 151.
——VI., 165.
——VII., 164, 166, 169, 171, 176, 177.
——VIII., 187, 200, 218.
Château-Gaillard, built by Richard I., and a proof of his talents as an engineer, 80.
Completed in one year, 81.
*Plan of castle and environs, 82.
Importance of the position, 83.
*Ground-plan of the castle, 85.
*View of part of the wall, 89.
*View of the keep, 91.
Siege of the castle, 92—94.
Machicolations of, 137.
Château du Bois, 73.
Château-Thierry, 77.
Châteaudun, donjon of, 105.
Châtillon-sur-Seine, 77.
Chauvigny in Poitou, 77, 98.
*Plan of the castle, 97.
Chemins de ronde, 25.
Chemise, or *wall of counter-guard*, 32, 98, 103, 105, 106, 108.
Cherbourg, siege of, 168, 169.
Circumvallation, the line of, 24, 93.
City, a Roman, 14, 15.
Clermont, Monseigneur, 169.
Coëhorn, M. de, 264.
Coitivi, Admiral de, 169.
Combourne, the castle of, 161.
Vicomte de, 162.
Compiègne, arbalétriers at, 165.
Condé in Brie, 114.
Constance, 123.
Constance, the lake of, 190.
Contravallation, the line of, 24, 29, 68, 93.
Corner-towers (*tours du coin*), 125.
Coucy, *plan of the town and castle, 76.
Donjon of, 105.
*Plans of the castle and keep, 106—110.
Section of the keep, 111.
Plan of construction, 112.
Hoarding at, 136.
Coucy, Enguerrand III. de, 113.
Counterguard, wall of, at Bonaguil, 179.
Counterscarp, the external slope of the ditch, as at Bonaguil, 179.
Cranequin, or *handle of crossbow*, 175.
Crécy, battle of, 78, 157, 166.
Crenelle (or *créneau*), 118, 139, 189, 249.
*With a wooden shutter, 118.
Crenellated parapets, 7.
Crèvecœur, 114.
Croix-Boissee, the boulevard of, 167.
Culverin, a small cannon, used in the sixteenth century, 207.
Curtain-walls, 61, 139, 141, 144, 145, 148, 176, 197, 206, 207, 208, 211.
*Plans of, 119, 120.
Cylinder, of wood, 250—252.

Dartz porcarissals, 45.
Derby, the Earl of, 158.
De Ville, Chevalier, 253, 261, 262, 264.
Deville, M. A., 80.
Dieppe, town walls, 77.
Discharging-arches, 208.
Domestic Architecture in England, Some Account of, 217.
Domfront, castle of, 98.
Donjon, the, 13, 31, 79, 80, 88, 90, 94, 96, 97, 98, 100, 104, 105, 106, 112, 151, 178, 179, 203, 221.
Douai, bastille of, 217.
Double palisades (*cadafalcs dobliers*), 45.
Drawbridges, *from a wooden tower, 65, 132—134.
At Carcassonne, 132.
Ducerceau, *Des plus excellents bâtiments de France*, 113.
Châteaux royaux en France, 133.

Du Guesclin, Bertrand, 155, 160, 162, 163.
Dunbar, the battle of, 217.
Duras, the town of, 171.
Durer, Albert, 227, 245.

Edward I., 217.
Edward III., 157—159, 165.
Embrasures (*fenestrals*), 45, 245—252.
Formed with gabions, 212.
Empire, the Roman, 3.
Engineers (*ingegneors*), 33.
Engines of war, 33—36, 64.
England, engineers of, 27.
English archers, 161.
Army in the fifteenth century, 161.
English, the, expelled from France by improved artillery, 169.
Enguerrand III. de Coucy, 105, 113.
Ermenville, Gérard d', 42.
Errard de Bar-le-Duc, 253.
Etampes, *plan of the keep, 99.
Evreux, corps collected at, 81.
*Ezekiel, figure of, from a MS. of the eleventh century, 26.

Falaise, 77, 98; *plan of the castle, 124.
Famars (Fanum Martis), 8, 9.
**Fascines*, 36.
Fausses braies, 199, 215, 229, 231, 259, 264.
Feudal castles, 79.
Horsemen (*gendarmerie*), 79.
Feudalism of France, 78.
Ferté-Aucoul, la, 114.
Flanders, supplies, provisions, and merchandize, 158.
Florence, 166, 219, 226.
Foot-soldier (*fantassin*), 78.
Forli, the fortress of, 254.
Fort, Guillaume, 38.
Forts, detached, use of, 28.
Fortalice (*châtelet*), 92, 94.
Fouque, Victor, *Recherches Hist. sur les Corpor. des Archers*, 166.
Fourneaux de mine, 259.
France, 27, 71, 72, 79, 245, 247, 265.
Feudal, 160.
France, the feudal nobility of, 162.
Francis I., 27, 151, 182, 187, 219, 237.
Frankfort-on-the-Maine, 233, 234.
Franks, the, 3, 24.
Free-archers (*francs-archers*), 166.
French army, 154, 155.
Froissart's Chronicles, 27, 135, 137, 141, 153, 155, 161, 171.

Gabions, 212, 213.
Gaillon-on-the-Seine, 81.
**Galleries*, plan of vaulted, 260.
Gallic tribes, 5.
Gallo-Romans, the, 2, 3, 23.
Garonne, the, 155.
Gascon army, 161.
Gate, the *prætorian*, *decumana*, *principalis dextra*, *principalis sinistra*, 7.
Gate, the, 33, 34.
Gateways, 125—131.
Gaul, 2, 3, 23.
Genoese infantry, 153.
Bowmen, 154, 160.
Revolt, 254, 255.
Géraut d'Aniort, 38.
Germanic customs, 3.
German infantry, 153.
Germans, the, 5, 22, 23.
Germany, 3, 27.
Gisors, the town of, 80, 83.
Godefroy, T., *Hist. d'Artus III.*, 167, 169.
Graveillant, a suburb of Carcassonne, 38.
Greek fire, 170.
Grille, or *iron railing*, 106, 117, 130.
Guiart, Guillaume, 34.
Guillaume des Ormes, 37, 38.
Guines, the Count of, 156.
———the comté of, 158.
Guise, town of, 77.
Guizot, M., 71.
Gunpowder, 27, 159.

Hainault, the Count of, 135, 136.
Ham, the castle of, 203.
Harecourt, the castle of, 170.
Harfleur, the siege of, 168.

Henry II. (of France), 151.
——IV., 261.
——the ship, 169.
Hero of Constantinople, 168.
Hides, or skins used to cover wooden ramparts, 5, *63, *65.
Hoards, or *hoarding*, vide *Hourds*.
Hooks (*falces murales*), 5.
Horn, or *pointed bastion*, 121, 122, 127.
*Plans of, 121.
Hôtel de Vauvert, 74.
Hourdés (coated with loam or mortar), 136.
Hourds, *hoarding*, or *hoards*, 58, 59, 60, 61, 62, 64, 99, 104, 110, 116, 118, 119, 120, 131, 136—138, 141, 244, 246, 247.
*Views of, 61, 63, 65, 129.
*Plan of, 137.
*Section of, 139.
*View of, from a MS. of Froissart, 141.
*View of, at Nuremberg, 247.
Hull, town of, 217.
*Part of the fortifications, 218.
Hyères, town of, 77.

Infantry, introduction of, 154.
Innspruck, town of, 223.
Issoudun, the treaty of, 80, 81.
Italian engineers, 237.
Wars, 259.
Italy, 27, 144, 197, 199, 218.

Jacques, or brigands, 161.
Jean, Maître, 170, 171.
John, King, 75.
—— the Good, 104, 161.
Joinville, describes the Greek fire, 170.
Jouy, gate of, at Provins, 122.

Kaiserstuhl-on-the-Rhine, 190.
Kas, *chaz-chateilz* (or moveable towers), 156.
Kingston-upon-Hull, 217.

La Hogue, port of, 157.
Langeais, castle of, 98.
Langres, Roman walls at, 15, 16, 182—188, 203.
Langres, *plan of the town walls, 182.
*Plans and sections of towers, 184, 185, 186.
Market-tower at, 187.
Languedoc, a brigand of, 161.
Laon, diocese of, 114.
Laval, Michauld de, 165.
Leghorn, bastille of, 217.
Limousin, the, 161.
**Lines of approach*, 63.
Lists (*lices*), 39.
Loches, town of, 77.
Castle of, 98.
*Plan of beaks at, 122.
Loire, the, 3.
**Loopholes*, in a Roman tower, 20.
*In a curtain-wall, 61.
*Arrangement of, at Carcassonne, 117, 118.
Lot (the river), 155.
Louis (Saint), 35, 37, 46, 49, 50, 58, 59.
——XI., 72, 114, 164, 171, 182, 200.
——XII., 254.
——XIII., 261.
Louvre (the), 73, 75, 76, 105.
Lubeck, city of, 217.
*Fortifications of, 219.

Machiavelli, 218, 226, 237, 253, 255, 256.
Machicoulis (*machicolation*), 61, 110—112, 125, 127, 128, 130, 137, 140, 141, 144, 148, 150, 179, 187, 189, 203, 221.
Maggi (Girol) *Della Fortif. della Citta*, 259.
Magna Charta, 78.
Maine, the river, 232.
Mangonels (*mangoniaux*), machines for throwing stones, 36, 38.
**Mantelets*, or *wooden shields*, 19, 36.
Marne, the river, 183.
Marseilles, siege of, described by Cæsar, 18—22, 33.
*View of bridge at, 223.
Martin, M. Th. Henri, *Morceaux du Texte d'Héron*, &c., 168.
Martinets, engines for hurling stones, 156.

Mats formed of cables, used by the Romans to protect wooden towers, 19.
Maximilian, the tomb of, 223.
Meaux, Viscount of, 114.
Mérian, *Topog. de la Gaule*, 226, 257.
Merlon, the solid part of a battlement, separated by the crenelles or openings : as at Bonaguil, 179.
Merovingian period, 24.
Metz, the city of, 257.
*View of the Mazelle gate at, 214.
*View of the barbican. 216.
*Bird's-eye view of part of the fortifications, 258.
Meudon, 73.
Meulan, town of, 77.
The dungeon of, 163.
Michelet's Hist. of France, Deposition of the Duc d'Alençon, 166.
Middle Ages, armies of the, 25.
Milan, the castle of, 219, 255.
*Bird's-eye view, 220.
Military Dictionary, &c., 207.
Mine, the, 37.
Mines and countermines, 259.
Monastery of St. Martin-of-the-Fields, 74.
Montargis, town of, 77.
*Plan of the castle, 95.
*View of the entrance, 133.
Montfort-l'Amaury, town of, 77.
——— Simon de, 10, 32, 33, 34, 43—45, 71.
Montlac, Blaise de, 205, 206.
Montmirail, 114.
Montrichard, castle of, 98.
Mottes, or *mounds*, 98.
Mouton (*langue d'oil*), a battering-ram, 25, 33.
Munstero, Sebast., *Della Cosmog. Universale*, 229, 232.
Musculus, or *rat*, 20, 21.

Narbonnaise, the, 3, 44.
Gate at Carcassonne, 39, 49, 51.
*Plans, 126, 128, 131.
*View, 129.
Narbonne, city of, 3, 235.
Newcastle-on-Tyne, 141.
Nieulay, the bridge of, 157, 158.
Nile, the, 35.
Nogent-le-Rotrou, 98.
Norman castles, 31.
Normans in England, 145.
Normandy, 81, 90, 157, 169.
——— the Duke of (afterwards King John), 155.
Nuremberg, 227, 228, 245, 249—251.
*View of parapet and hoarding at, 247.
*Embrasure of the Laufer gate at, 250, 251.
Nuys, fortifications of, 201, 202.

Oisy, 114.
Orange, town of, 77, 145, 200.
*View of part of the fortifications, 201.
Oriental projectiles, 30.
Orillon, or *oblong bastion*, *view of, 235.
*Plan of, 236.
Orleans, city of, 95, 167.
Siege of, 170, 171.
Osier parapets, 5.
Outworks, *antemuralia*, *procastria*, 7.

Pacy-sur-Eure, 81.
**Palisades*, at Carcassonne, 12.
Attacked by battering-rams, 25.
Use of, 27, *41.
Parapets (*chemins-de-ronde*), 208, 209, 221.
Paris, 73—75, 145, 156, 157.
*Plan of, in the thirteenth century, 74.
*In the fourteenth, 75.
Imperial Library of, 25, 141, 171, 351.
Parker, Mr. J. H., 217.
Pavia, siege of, 27.
Périgueux, *plan and *view of tower at, 188, 189.
Petraria, an engine for throwing stones, 38, 39, 43.
Philip Augustus, 34, 73, 80, 81, 90, 92—94, 114, 167.
——the Bold, 47, 50, 123, 127, 132.

Philip de Valois, 72, 151, 155, 158, 159, 161, 162.
Picardy, march of the English through, 157.
Pierrefonds, the castle of, 141—144.
*View of part of the castle, 142.
*The same restored, 143.
Platform, or *cavalier*, 222—225.
Pockets (*fourneaux*), 259.
Poitiers, the city of, 15, 78.
Portcullis, 127, 128, 129, 131.
Provençe, the people of, 43.
Provins, the keep of, 99.
*Plans of the keep, 100, 101.
*Elevation, 102.
*Section, 103.
Paris gate at, 103.
Gate of Jouy at, 122.
Gate of St. John at, 122.
Puilaurnes, Guillaume de, 10.

Radulfus, Bishop, 46.
Ramparts, 209—211.
Wooden, 4—6.
Rat, the (*musculus*), 20, 21, 33.
Ravelins, 243.
*Plans of, 244.
Raymond, Count, 45.
Remparer, to, 209.
Return, a (*redent*), 127, *248.
Rheims, the city of, 114.
Rhine, the, 190.
Rhodez, gate, 40.
Ribaudequins, 176.
Richard Cœur-de-Lion, his military genius and talent as an engineer shewn in the Château-Gaillard, 80, 87, 94, 137.
Robert, King, 73, 74.
Roger de Lacy, 92—94.
Roman entrenched camps, 6—8.
*Wooden ramparts, 6.
Wooden towers, 8.
*Method of constructing walls, 9.
*Testudo, or tortoise, 6.
Fortifications, 9, 17.
*Plan of walls, 17.
*Section of tower, 20.
Roman soldiery, 5, 6, 21, 22.
Towns, 14.
Towers, 19, 20.
Romans, the, 3, 13, 17, 18, 19, 24, 29.
Rome, 3, 5, 15.
Roucy, Count of, 114.
Rouen, 81, 83, 90, 93, 168, 235

Saint Ladre d'Orléans, 167.
—— Louis, 123.
—— Omer, 158.
—— Pol, the Constable of, 203.
—— Sernin, 46.
Salisbury, the Earl of, 170.
Sangattes, the mount of, 158.
San-Giorgio, 219, 226.
San Michele, 237.
Saracens, the, 35.
Saracen wall, 40.
Saumur, the town of, 77.
Scala, Giov., *Delle Fortif.*, 240, 241.
Scaling ladders (*échelades*), 146.
Schaffhausen, fortress of, 203, 245.
*Fortifications of the bridge, 191.
*Plan of the citadel, 192.
*View of one of the bastions, 193.
*Plan of the bastion, 194.
*Plan and section of embrasures, 195.
*Plan of the platform, 196.
*Bird's-eye view of the fortress, 198.
Chemin-de-ronde, 191, 192.
Schayes, *Hist. de l'Architecture en Belgique*, 8.
Seine, the, 73, 81, 83, 84, 92, 94.
Septimania, 3.
Shields (*pavois*), 150.
**Shutters*, wooden, 118, 134.
Sienna, defence of, 205.
*View of part of the fortifications, 206.
Spiral ramp (or *inclined way*), 191, 195.
Springald, a kind of sling for throwing stones, 158. (The same name was afterwards applied to a kind of cannon.)
Stein-on-the-Rhine, 190.
Stone merlons, 204, 205.
Sylvestre, *Vue des Maisons Royales et Villes*, 151.

Talus, or *incline*, 204.
Tancarville, the Count of, 156.
* *Tapecu*, or *shutter*, 134, 135.
Temple, the, 74, 75.
Tenailles, 244.
Termes, Olivier de, 38.
* *Testudo*, the Roman, 6.
Timber hoards, or *hoarding*, 136—139, 141, 149.
Toëni-on-the-Seine, 92.
Topographie de la Gaule, 182.
* *Tortoise*, the (*testudo*), 5, 6, 168.
Toulouse, ramparts of the Visigoths, 3.
  Besieged by Simon de Montfort, 44.
  Defended chiefly by timber-works, 45, 71.
  Chateau Narbonnais at, 9.
  Place du Salin, 10.
Tour-de-ronde, 102.
Tournay, the men of, 158.
Towers, the Roman mode of building, 19.
  Section of one in construction, 20.
  Of wood covered with skins, at Bourges, 5.
Towers on rollers (*baffraiz*), 33.
Trajan's column, 4, 6, 8.
Traverses, 225.
Trebonius, C., 18.
Treitzsaurwen, Mark, *Le Roi Sage*, 211.
Trencavel, 37, 46, 47.
Trenches, 36, 167, 168.
  With gabions, 213.
Trésau, Tour du, 49, 54.
Troyes, fortifications of, repaired in 1541, 237.
  *Plan of one of the bastions, 238.
Tubs, or *semals*, 45.
Turkish engines, 31.
—— *Petraria*, 38.

Vauban, 253, 257, 261, 264.
Vegetius, 7, 10.
Vernon-on-the-Seine, 81, 83.
Verona, fortifications of, 223.
  *View of a cavalier on a bastion at, 224.
  Fortified by San Michele, 237.
Vexin territory, the, 80.
Villeneuve d'Agen, 177.
———— le-Roi, 122.
*Vincennes, 133.
  Plan of the castle, 152.
Visigoths in the fifth century, extent of their dominion, 3.
  Fortifications of the, at Carcassonne, 10.
  Towns of, 16.
  *Plan of town, 17.
  Walls of town, 37.
  Walls repaired, 47.
  Towers of, 50.
Vitruvius, 8.
Voisin, Pierre de, 42.

Wallia, 3.
William the Breton, 90, 94.
Wooden **ramparts*, protected by hides, 5.
  Scaffoldings, 6, 45, 71, see also *Hourds*.
  *Roman work, 4.
  German work, 6.
  *Towers on Roman walls, 8.
  *Palisades, at Carcassonne, 12.
  *Defences of a breach, 41.
  *Hoarding, 61.
  *Plan of, 137.
  *Door of a bastion, 67.
  *Shutter, hanging, 12, 118.
    Suspended, 134.
    *On a pivot, 134.
  *Ramparts, 211.